NAPOLEON

BY MATTHEW FLETCHER

BOOK PRODUCTION

Cover Art: Chris Collingwood, used with permission.

Internal Art: Richard Knötel used with permission. Alan & Michael Perry.

Editing: Keith Pinfold, Martin Buck, David Sheppard & Matt Keefe.

Additional Material: Kevin Dallimore, David Sheppard, Keith Oliver, Alan Perry & Matt Keefe.

Play-testing: Matthew Fletcher, Bernard Lewis, Ian Miller, David Cross, Che Webster, Graeme Basnett, Robert Townsend, Andrew Chesney, & John Carter.

Layout & Graphic Design: Kevin Dallimore & Keith Pinfold.

Photography: Kevin Dallimore & Tom Weiss.

Painting Guides: Kevin Dallimore & Matthew Fletcher.

Special thanks to

Keith Oliver of the Napoleonic Association (www.napoleonicassociation.org) Stephen Maughan & Michael Leventhal of Greenhill Books.

MINIATURES & SCENERY

All miniatures produced by: Foundry Miniatures Ltd. Visit www.wargamesfoundry.com

Miniature Painting: Matthew Fletcher, Ian Miller, Ian Clark, Paul Ballard, Jez Griffin, Jez Farminer, David Sheppard, Lee Clapham, Paul Cubin, Aaron Ishmael, James Bowen, Tom Weiss, Wayne John, Ian Stables, Martin Buck & Kevin Dallimore.

Miniatures from the collections of: Kevin Dallimore & Matthew Fletcher.

Scenery from the collections of: Kevin Dallimore & Matthew Fletcher.

First published in Great Britain in 2009

By Foundry Publications, 24-34 St Mark's Street, NG3 1DE, United Kingdom. Tel 0115 841 3000.

www.wargamesfoundry.com.

ISBN 10 1-901543-18-8

ISBN 13 978-1-901543-18-6

www.wargamesfoundry.com

Napoleon bids farewell to the Army on his abdication in 1814.

CONTENTS

Austrian Infantryman

CONTENTS

French Infantry Officer.

FOREWORD

There are only two forces in the world, the sword and the spirit. In the long run the sword will always be conquered by the spirit.

— Napoleon Bonaparte

I first met Matt Fletcher when he was employed by a well known fantasy games company based in Nottingham; he was an 18 year old, wide eyed, fresh faced teenager looking like he had found Aladdin's Cave as he scanned all of the models, games and books stacked on the warehouse shelves. Two of Matt's childhood dreams had been realised; he was not only involved in a rapidly growing company making and selling the products he loved, he was also heavily involved with the team responsible for the play testing and development of games and the writing of books. Everyone was impressed by Matt's energy, enthusiasm and knowledge; he had a great knowledge of all things fantasy and, as I was later to find out, an even greater knowledge of all things historical.

Matt soon realised that these new experiences might lead to the fulfilment of yet another childhood dream; to write a book that combined his extensive historical knowledge of all things Napoleonic with the new experiences gained in the creation and development of books, games and game mechanics. Little did he realise how much time it would take for this dream to be fulfilled.

Eighteen years have since passed and I think Matt has been very surprised (and sometimes frustrated) with the length of time that it has taken to produce the book that you now hold; not only for the time it has taken him to produce and structure a script suitable for submission, but also for the time it has taken for his script to be edited and augmented to reach the high standard expected from any Foundry publication.

If you are interested in painting, collecting or wargaming with toy soldiers, or interested in the Napoleonic era any way, then this book is for you; it has many articles that set the scene and mood for ensuing battles by detailing the historical background and build up to the many Napoleonic wars and campaigns that took place. For the miniature painters and collectors, Kevin Dallimore has added his own article on painting Napoleonic models using his unique and world renowned three colour method. For the wargamer, Matt has created a truly unique set of rules that have been amended and developed many times over the years to provide an evening of fun and entertainment; he has also written his own article on how to speed paint an army to a standard that any wargamer would be proud of.

Matt is very proud of this book and wants everyone who reads it to experience the great fun and pleasure he gets from painting and playing games with toy soldiers.

For Matt, the book you now hold is another dream come true.

Keith Pinfold
Foundry Publications

French Chasseurs a Cheval of the Guard.

AUTHOR'S FOREWORD

A man like me troubles himself little about a million men.

— Napoleon Bonaparte

My introduction to the world of gaming with miniature models came at the age of eleven when, along with a group of school friends, I discovered *Dungeons and Dragons*. I was instantly hooked and soon every spare moment was filled with battling goblins and dragons and with attempts to paint the few miniatures my pocket money could afford. The local gaming store soon became a second home!

I had no idea that a more expansive world of wargaming existed beyond the cramped tunnels and small skirmishes of our dungeon adventures until one hot summer day when I was invited round to the house of my friend Stuart Tilley. His father, Robert, was a keen wargamer who had noticed our interest and decided it was time to introduce us to gaming on a larger scale.

I walked into the lounge to be confronted by the largest dining table I had ever seen. It was covered with a green cloth, scattered with model trees and buildings and hundreds of beautifully painted Airfix Napoleonic miniatures. The effect on me was immediate and lasting. I knew nothing of the Napoleonic era but I was awestruck by the sheer scale and numbers of the massed ranks of soldiers in their bright uniforms; once again I was hooked! Without any idea of who the Prussians were or why we were fighting them Stuart and I took command of the French army and eventually achieved victory.

I was lucky to have a mentor like Robert Tilley. With his help and advice I was soon painting endless ranks of plastic models, making terrain and poring my way through countless books on the Napoleonic period. *The Campaigns of Napoleon* by David Chandler and *Uniforms of Waterloo* by Philip Haythornthwaite soon became the staple reading material of my teenage years. I may not have been able to say which song was currently number one in the charts, but I could name the colour of the uniform facings of the 2nd Hussars without a second thought! My infatuation with the period has never died. I still read all the books I can find on the subject and still try, whenever possible, to paint two units of Napoleonic troops every month. The French army has always been for me the epitome of the period, capturing perfectly the style and flavour of the era. I confess that I am a Francophile at heart!

It was fairly inevitable that after so many years of gaming with other people's rules I would eventually devise a set of my own. I became frustrated with the similarity and predictability of many rules and wanted a game that recreated not only the chaos and confusion of battle, but also enabled the players to profit from being a capable Commander in Chief or General, or suffer the consequences of being a poor one! Most of all I wanted to recall the fun and excitement of those games that started it all, played across a green cloth on a dining table in Nottingham.

These rules have evolved many times over the last ten years, and in that time I have had invaluable help, suggestions and advice from Che Webster, John Carter, Dave Cross, Bernard Lewis, Ronnie Shilton, David Sheppard and Ian Miller; my apologies to anyone I may have forgotten.

Napoleon mounted on one of his Arab horses inspects his Imperial Guard Grenadiers.

Introduction

I love a brave soldier who has undergone the baptism of fire

— Napoleon Bonaparte

Marshal Ney turned in his saddle to regard the troops arrayed behind. As his eyes passed along their familiar ranks he lost himself momentarily to pride; these fine men, these five thousand elite Cuirassiers were his to command. They would follow him anywhere, advance at his order, charge with valour and fight like lions. And he was their leader, bold, charismatic, the bravest lion of all.

His gaze turned back to the field ahead. Today he would be victorious. His charge would crash upon the enemy like a wave in a tempest carrying the day and bathing him in glory. Today he would return the hero and stamp out once and for all the rumours that two days ago at Quatre Bras he had failed, that he had been weak. The rumours were wrong, Ney did not know the meaning of failure.

The artillery batteries suddenly ceased their bombardment. Now was the moment. The horsemen advanced in echelon, their pace quickening as they drew closer to the enemy. Spurred on by adrenalin and the heady scent of destiny Ney sprang forward in his stirrups slicing his bright sabre to and fro through the heavy air. The sodden earth shuddered to the hammering of hooves as five thousand horses broke from a canter into a full gallop behind him.

It was the afternoon of the 18th June 1815. It was Waterloo.

"Vive l'Empereur!"

Who has not wanted to be Marshal Ney, the 'bravest of the brave', leading five thousand cavalry to glory over Wellington's army? Or Wellington or Blücher or even Napoleon himself, rewriting history by achieving victory on the battlefield of Waterloo?

Well now you can; these wargames rules will enable you to recreate the conflicts of the Napoleonic era, from relatively small-scale skirmishes involving a few regiments to huge pitched battles where vast armies roll across the landscape.

These rules place you in the position of the Commander in Chief of your army. You will benefit from your ability to understand the strengths and weaknesses of the soldiers under your command, to know how the troops will respond to your orders in the heat of battle and from your prowess as a leader in developing a battle strategy that will enable them to perform to their full potential.

Great value is placed on your ability to make on-the-spot judgements as the fighting unfolds, to make real tactical decisions that will directly affect the outcome of the battle. Troop deployments and the timing of advances and attacks are crucial to your success.

There are no endless charts and tables of statistics to study; incisive, intuitive and creative leadership in the heat of combat are the key elements that will lead you to victory.

These rules will provide you with an evening's gaming that is dramatic, tense and, above all, fun. The rules are simple to play, but they are not simplistic, they are easy to learn yet a challenge to fully master.

May the best Commander win!

Europe in 1813.

British Infantryman.

THINGS YOU WILL NEED

If the art of war were nothing but the art of avoiding risks, glory would become the prey of mediocre minds. I have made all the calculations, fate will do the rest.

— Napoleon Bonaparte

THE ARMY

You are the Commander in Chief of a Napoleonic army. During the game you will be directing the movement and actions of armies of miniature soldiers across a tabletop battlefield. Before you can begin, you'll need an army!

These rules are designed for use with miniature soldiers about an inch tall (usually described by model manufacturers as 25mm or 28mm models) grouped together with others of the same type to create fighting units. We prefer and recommend miniatures of this size as they are easier to paint well and have a more tactile feel on the tabletop than smaller models. All the pictures of miniatures in this book are of this size.

If you have an existing army of tabletop veterans, then dust them down and send them straight into battle. If not, the information on the next page will give you a good idea how to go about choosing your army and assembling it ready for battle.

TAPE MEASURE

As well as your army, you will need a tape measure marked in inches (") as these rules use inches to measure distances on the tabletop. If you want to play these rules using a metric tape measure, simply multiply the distance given in inches by two to represent the number of centimetres to be moved. We think this is a simple and more satisfactory measurement ratio and much better than trying to use fiddly measurements like half a centimetre and so on. To restrict all bloodshed to the tabletop, it is highly recommended that all players agree before the battle commences which measuring system is to be used.

DICE

You will need at least eight six-sided dice and each dice is referred to in the rules as a D6: a 1D6 means you should roll one dice, 2D6 means you should roll two dice and so on.

Occasionally you will also be asked to roll a D3 (written as 1D3, 2D3 and so on). Don't worry, you don't need to go looking for a three sided dice – there's no such thing! Instead, simply roll a D6 and halve the score, rounding odd fractions up. So, a score of 1 or 2 = 1, a score of 3 or 4 = 2, and a score of 5 or 6 = 3, giving you results from 1 to 3, hence a 'D3'.

PLAYING SURFACE

Napoleonic battles were large affairs, and if you're planning on fighting one out on the tabletop you'll need a lot of space! A space of about 8' x 4' is ideal. You can manage with 6' x 4', but any less than this and the battlefield will look a little crowded. A sheet of board or green cloth laid down on top of a dining table should suffice for most games.

THE GENERAL'S PROVISIONS

A ready supply of food and drinks is also advisable to keep you on top form throughout the game. Experienced players find that a good bottle of *vin rouge* is highly in keeping with the flavour of the period as French armies drank quite a lot of wine before going into battle!

An army marches on its stomach.

— Napoleon Bonaparte

French Young Guard, an often under represented part of Napoleons Imperial Guard.

An army's effectiveness depends on its size, training, experience, and morale, and morale is worth more than any of the other factors combined.

— Napoleon Bonaparte

French Carabinier (Knötel).

French Horse Grenadier Trumpeter.

The Army

Two armies are two bodies which meet and try to frighten each other.

— Napoleon Bonaparte

Battles are fought out between armies of miniature soldiers, representing tens (or even hundreds) of thousands of men. Selecting and preparing your army for battle is the single most important part of the wargaming hobby. Not only is there the question of choosing your forces but also of preparing them for battle – painting those many miniatures, basing them and assembling them into units. Whether you are a new wargamer or a hardened veteran, it is recommended that you read this section carefully; while some of the contents may seem familiar, this section also introduces important themes and concepts central to the rules. A thorough understanding of them will make digesting what comes later that much easier.

The Army section is divided into four parts:

Collecting an Army

How to choose which army to collect as well as how to go about choosing your miniatures.

Prussian Hussar (Knötel).

French Chasseur a Cheval (Knötel).

Armies of the Napoleonic Wars

An overview of the historical structure and organisation of the armies of Europe in the time of Napoleon. This part also describes how these historical formations are represented on the tabletop.

Units

Here we present information on base sizes, the number of miniatures in each unit and how they are grouped together.

Formations

The formations your units may take on the battlefield; the shape in which they go to battle.

If you are new to tabletop gaming and are about to collect your first army, or even if you are an experienced wargamer trying out a new period, the wealth of options open to the Napoleonic wargamer can be overwhelming.

CHOOSING AN ARMY AND PREPARING IT FOR BATTLE

WHICH ARMY?

First of all, which army do you want to collect? Which of Europe's proud nations do you want your forces to hail from? For most the answer to this question lies in gut instinct. Perhaps you are inspired by a particularly colourful and inspirational Commander in Chief like Marshal Ney, Wellington or Blücher. Maybe your choice of army will be swayed by patriotism or the reputation and track record of a particular nation's troops. You might even choose your army based on the fact that you like their uniforms and colours more than anyone else's. Your enthusiasm for a particular uniform can become an important factor when you are painting it over and over again! Literary works, paintings, sculptures, movies, all these factors can sway and inspire your choice of army – it is always very rewarding to collect, paint and game with an army for which you have a passion.

There are a great many armies to choose from, the most famous of which are, of course, the British, the French, the Austrians, the Prussians, the Russians. Take a look at the pictures in this book and see which army particularly takes your fancy.

FRENCH

AUSTRIAN

PRUSSIAN

BRITISH

RUSSIAN

You must not fight too often with one enemy, or you will teach him all your art of war
— Napoleon Bonaparte

Prussian ADC.

French infantry divisions, the units have staggered or echeloned their attack for maximum effect, notice how the light infantry unit (Chasseurs) are at the foremost of the units, deploying a screen of skirmishers.

An allied army of Russian, Prussians and Austrians advance against the awaiting French. Coalitions of different nations frequently fought against the French.

French Infantryman.

Austrians surge forward! These units and others in this book are using slightly different unit model numbers and bases sizes, this makes no difference in these rules, so if you have an old established army there is no need for any rebasing!

A Prussian infantry division with foot artillery support, a mixture of regular, reserve and militia (Landwehr) advance in a checker board deployment; this allows space on the battlefield for redeployment.

Prussian Reservist Infantryman.

HOW MANY MODELS?

Once you have chosen which army to collect you will have to work out which miniatures are needed to represent it. This book contains an Army List section detailing the units and troop types that made up many armies of the period and this will provide a useful guide in collecting your miniatures. Most armies will be a mix of infantry, cavalry and perhaps artillery and a typical army might consist of a couple of divisions each of three line infantry units, a light infantry unit and a Commander in Chief. Some of the army lists then allow you to add an artillery battery, or additional infantry and a unit of light cavalry. This will give you enough of a collection to start fighting battles using these rules. To make this part of the process easier you can always buy ready-designed starter armies, which give you a head start and can often save you money. As you become more experienced with the rules you can add more units, including elite troops, and gradually watch your army grow into something substantial that you can be really proud of on the battlefield.

HISTORICAL ARMIES

You can collect an army based on a particular battle or famous division and finding the relevant historical army lists to determine troop types, uniform and colours is easy. Historical army lists may not be the same as some of the lists given in this book, however, as the latter have been designed to utilise specific game rules.

The army lists in this book are designed to give an accurate representation of the mix of troops found within each army, though they don't seek to mimic the formations of these armies precisely. We've included an overview of the structure and organisation of the armies of the period on the next page but don't let this information overwhelm. For now, all that's really important is that you understand the parts that go to make up a wargames army.

Napoleon

The British army deployed in its favoured line formation.

British Infantry Sergeant.

French Lancers.

ARMY LIST EXAMPLE

To give you an idea of a basic starter army, below is a sample list for a small British force that will provide enough troops for an evening's gaming across an eight feet by four feet table. Note that the force consists of two divisions with additional artillery and cavalry and that each of the units has been assigned a points value - this is so that players can each keep track of the size of their armies. It is important to realise that, before starting to play, opposing armies should have approximately the same points value for a fair and more balanced game.

First Division

Unit	**Points Value**
Line Infantry Regiment	20
Elite Line Infantry	25
Highlander Regiment	25
Rifle Detachment	10
	80

Second Division

Unit	**Points Value**
Line Infantry Regiment	20
Line Infantry Regiment	20
Light Infantry Regiment	25
Rifle Detachment	10
	75
Foot Artillery Battery	**60**
Light Cavalry	**35**
Total Points	**250**

PAINTING YOUR ARMY

If you are creating a new army then you will want to spend some time painting your troops and making them look as resplendent as possible for the battles ahead. Painting your army is as important and as enjoyable as playing the game itself. Some people copy the actual historical colours of the troops at the time – which can in itself be the subject of much discussion and debate – whilst others really personalise their armies and create their own colour schemes and flags. There is a useful painting guide on page 92 of this book to give you a few tips and ideas on how to make your army look magnificent on the battlefield.

UNIFORMS

Like painting and modelling, the study of uniforms can be an absorbing pastime in its own right and is just one aspect of the many facetted wargaming hobby. The complexity of Napoleonic uniforms with their precise etiquette and exotic names like colpack and pelisse for example, can seem daunting at first, but persevere and they will become second nature to you in no time at all. Most armies of the period were dressed and equipped in a similar way and sported the latest military fashions, but clothing generally became simpler and more practical as armies grew in size toward the end of the period. The mass of campaign troops were not clothed in the beautiful full dress regalia of uniform prints at all but in tatty, discoloured rags and voluminous greatcoats.

UNIFORM COLOURS AND DETAILS

There are numerous reference books available for Napoleonic uniforms and many online resources display cards and prints detailing the garb of the great powers of the day. Most military museums exhibit original uniforms, so you can get a first hand look at the depth of colour and complexity of detail involved. Napoleonic re-enactors are another useful source of reference as they go to great lengths to create authentic clothing and kit. You will have to look a bit harder to find references for some of the minor armies though and will notice that sometimes, for all sorts of reasons, sources contradict each other. In some cases you have to resort to your own judgement on how a uniform will be depicted and on whether you opt for the 'bright and shiny' or 'war torn and weather-beaten' look for your troops.

French Line Lancers attack Prussian Infantry

Armies of the Napoleonic Wars

An army ought to only have one line of operation. This should be preserved with care, and never abandoned but in the last extremity.

— Napoleon Bonaparte

European armies, over time, copied the structure of the Napoleonic army, therefore success in battle became of function of strategy and numerical superiority. By describing the structure of the Napoleonic army we are, in effect, describing the structure of many of the armies that fought during this period.

Napoleonic armies were complex entities built along strict logistical lines and were formed by a combination of battalions, squadrons, regiments, brigades, divisions and companies.

A battalion was made up of a number of companies.

A number of infantry battalions grouped together formed a regiment.

A collection of infantry regiments in turn made up an infantry brigade. Cavalry brigades likewise consisted of regiments which were made up of a number of smaller squadrons.

Divisions contained several brigades and would often have artillery attached in support.

Corps was constructed from a collection of infantry divisions plus cavalry brigades or divisions. They also contained pioneers, engineers and other ancillary units and were, in effect, small independent armies that were capable of holding their own even against larger enemy forces.

The French corps system was so effective that before long most armies were fielding their troops in the same manner. Corps marched in support of each other, with less than a day's march between them. This meant that reinforcements were never too far away. Often Napoleon's enemies were unsure just how many troops were coming up against them; a supporting corps would arrive unannounced onto the battlefield and quickly dilute any numerical advantage Napoleon's enemies thought they had. To make matters worse for the enemy, the reinforcements had the disconcerting habit of arriving in the flank or even the rear of their troops.

Many European Armies often held reserves of troops; crack formations of hardened veterans selected from the best recruits, or hand picked from other units for their heroic deeds. Austria and Russia had massive Grenadier divisions, while Prussia, Russia and Britain fielded battle hardened Guard formations.

None of these had such a fearsome reputation – or caused as much panic among their enemies – as Napoleon's Old Guard. Even amidst the carnage towards the end of the Battle of Waterloo, when everyone realised the French were defeated, The Old Guard would not lay down their arms and surrender. Over many years the French Imperial Guard expanded to such a degree it became an army in itself.

Russian Cuirassiers, soldiers from various units to demonstrate their different regimental "facing colours" this helped General's to identify particular units within their army.

French Divisional Structure

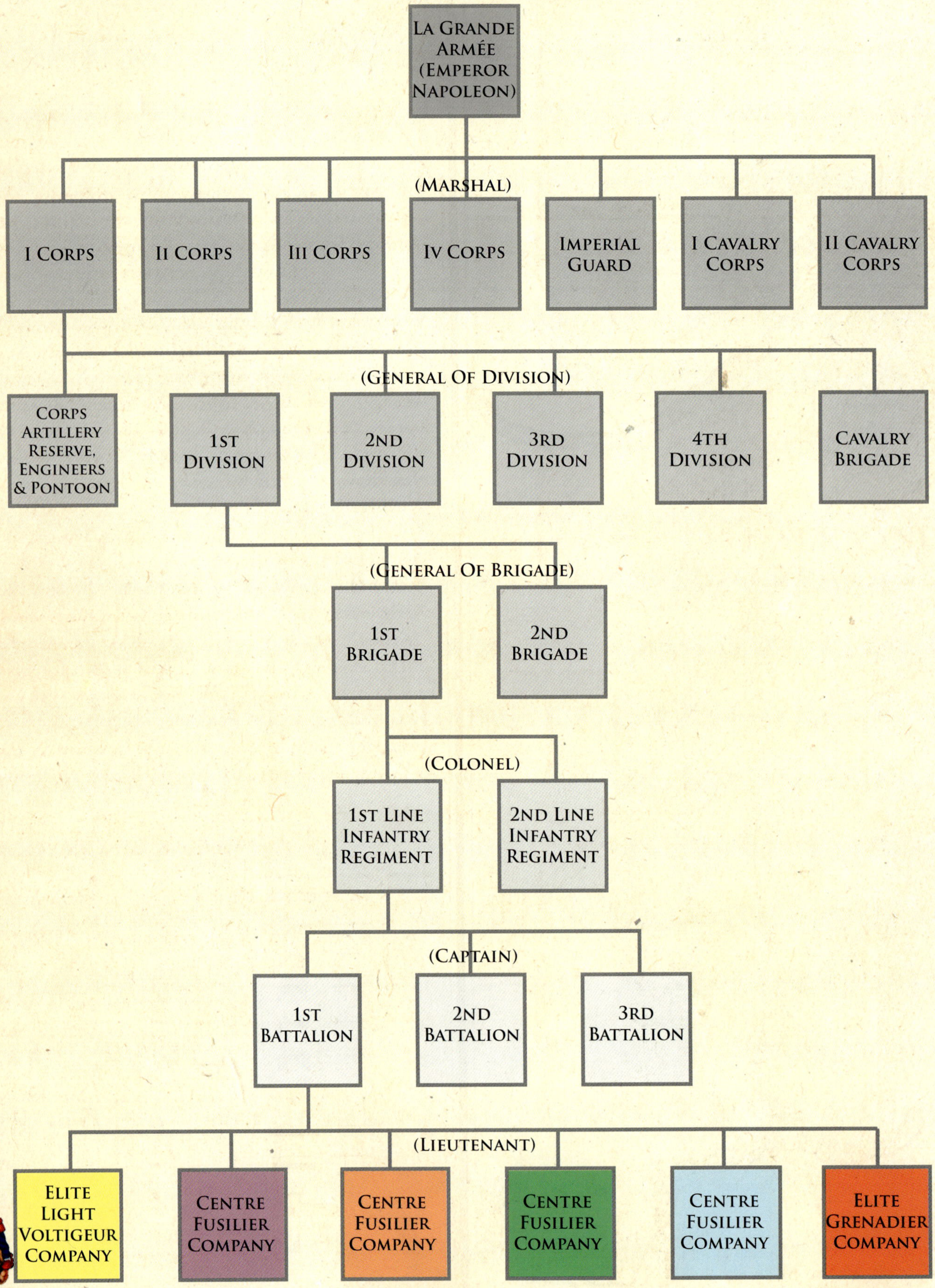

French Pioneer.

A Structure for Playing the Game

As you can imagine, it would be very difficult to represent these many different army sub-divisions on the gaming table. It is also somewhat unnecessary, since smaller sub-divisions like battalions, squadrons and regiments seldom acted alone anyway. Instead, in the game we use just two sub-divisions for the vast majority of fighting bodies: **units** and **divisions**. In addition, in the case of cavalry, there is also the **brigade**.

A unit

Divisions are groups of three or more infantry or cavalry units. Each division can be thought of as representing a brigade of a division, or in very large games even a corps. Divisions can include detachments of riflemen and artillery

A division

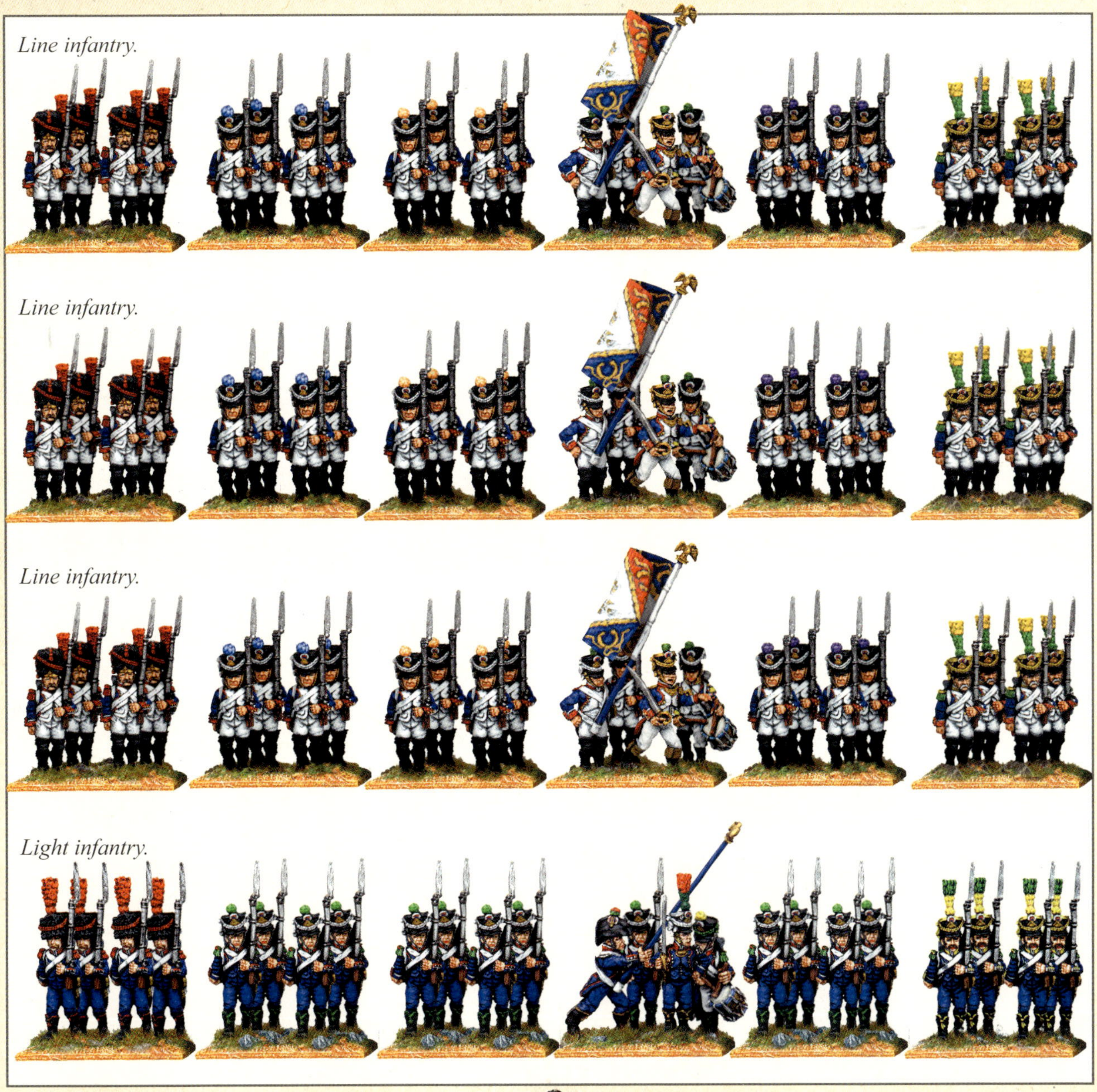

A cavalry brigade consists of one or two units of cavalry. These are more often used in support of other divisions in the army and note that cavalry units may also be included within other divisions.

A French light cavalry brigade, consisting of Chasseurs a Cheval and Line Lancer units.

Your miniatures will be grouped into units and a number of units will be grouped together into a division. An army is made up of a number of divisions. Full details can be found in the Army Lists on page 115-116.

A French and Bavarian infantry division advance in a mixed column formation.

Representing Historical Formations

In terms of historical accuracy, although these game divisions may not genuinely represent historical divisions and not all the lists provided are based on a particular historical formation, they do typify the types of unit and unit strengths that existed in the various campaigns of the Napoleonic Wars.

When refighting specific historical battles you might need to modify the Army Lists in these rules to suit the battle. The use of 'units' to represent army sub-divisions is fairly fluid and can be upgraded or downgraded accordingly; so if you want to fight a large battle on an epic scale, rather than using hundreds or thousands of models, you could treat each wargames division as a corps instead.

This way you can stage huge battles using just twenty infantry units a side and, as such, could effectively field the whole Grande Armée if you wanted to!

For smaller scale actions, such as the skirmishes of the Peninsular War, each wargame division can be used to represent a brigade.

There are various types of unit which go to make up divisions and brigades. These are detailed on the next pages.

Units

Soldiers generally win battles; generals get credit for them.

— Napoleon Bonaparte

Base Sizes

Although these rules do recommend (and refer to) base sizes for 28mm models, if your existing army is based differently there is no need to re-base them as these rules will work fine with any base size as long as all players have based their armies in a similar manner. For more information on using other base sizes see the appendices on page 212.

Infantry Units

In these rules each model of an infantryman represents thirty to forty actual men and should typically be deployed in units consisting of twenty to twenty-four models. The Army Lists give more detail on the appropriate number of models required per unit depending on nationality and period.

To increase the stability of your models, and to make them easy to move around the battlefield, base them two models wide and two models deep on some card, plastic or plywood.; each base of four infantry models should be 40mm square.

Example: A French line infantry battalion of the Napoleonic Wars consisted of four Fusilier companies, one Grenadier company and one Voltigeur (light infantry) company. This can be represented as a wargames unit by equating each company to a base of four models. The twenty-four model unit would thus have four bases of Fusiliers (sixteen Fusilier models), one base of Grenadiers (four Grenadier models) and one base of Voltigeurs (four Voltigeur models).

The way different companies within a unit are represented on the tabletop is really up to you as it has very little impact on the rules in terms of game play. Dividing infantry units up into Fusiliers, Grenadiers and light troops does make a difference visually though, as the unit looks more interesting and colourful on the battlefield. Grenadier and light companies were often called flank companies as they were deployed at either end of the parent unit; (the vibrant additions and colours they had on their uniforms can become an interesting painting project).

Victory belongs to the most persevering

— Napoleon Bonaparte

Infantry base and an overhead plan with dimensions

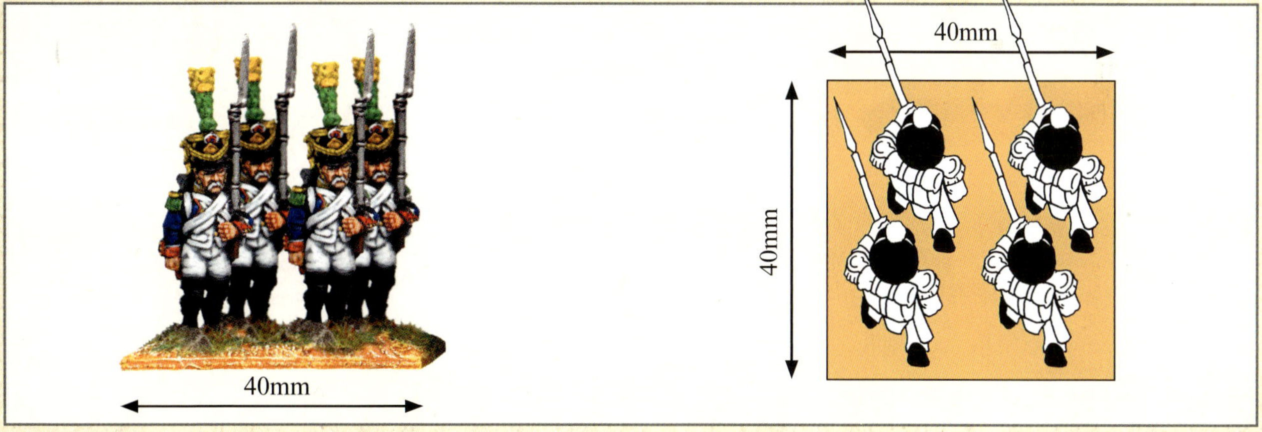

French Infantry unit, consisting only of fusilier models.

French Infantry unit, consisting of a mix of troops, grenadiers, fusiliers (command) and Voltigeurs (light infantry).

To further enhance your infantry units you can include an officer model on one of the central bases (sometimes mounted, as this looks impressive) plus a drummer and standard bearer. You should always consider adding an extra officer, ensign and/or musician model as you see fit as they look good on the table and can provide a welcome respite from painting rank and file models!

The Advanced Rules allow for light infantry to be deployed as skirmishers. To simulate their irregular nature skirmishers should only have two models per 40mm square base arranged in a non-regimented fashion. To enhance the irregular look of skirmishers you can place the occasional model individually on a base this size if you want. Light infantry offer particularly rich modelling opportunities – miniatures kneeling to fire, running, occupying cover and so on, all of these positions can be modelled onto skirmisher bases.

French Light Infantry.

Napoleon at the Battle of Austerlitz.

The Commander in Chief

You are the Commander in Chief, the commander of your army and are represented on the tabletop by either a single mounted model or a group of models arranged on an irregular or circular shaped base. As the Commander in Chief you will be directing troop movements, deciding tactics and ordering troops to attack.

When attaching a base to the Commander in Chief it is worth remembering that it will be the focal point for the army, so it is worth spending some extra time with your painting in an effort to make it look impressive – after all, it is supposed to be you! You could try adding some Aide-de-Camps, Sub-Commander in Chiefs, or even the occasional mistress to your base to add some interest.

The Commander in Chief plays a key role in maintaining discipline and morale and you will discover, as you read through the main rules, that it is crucial to be as mobile as possible to ensure that you are able to control the battle where the fighting is at its thickest.

Each division and brigade is directly controlled by its own Commander, who is subordinate to the Commander in Chief, and must stay close to his division or brigade at all times.

This may be slightly confusing, but bear in mind that the Commander in Chief is mobile and has to move to many areas on the battlefield to direct and give orders to his subordinate Commanders, whereas the subordinate Commanders remain with the brigade or division they are commanding and move with them.

During the Napoleonic period Commander in Chiefs still led from the front and were often to be found in the thick of battle – legend has it that Marshal Ney was the last Frenchman to retreat from Russia!

With Commander in Chiefs in such close proximity to combat it was not uncommon for them to become casualties and this is dealt with in the Advanced Rules section. For the Basic Rules you just need to remember that Generals do not personally fight and are not affected by any special rules for terrain or engagement ranges.

Any command-based measurements that need to be made to the Commander in Chief and Generals are made to the model himself and not to the edge of his base.

Cavalry Units

As with infantry, cavalry models represent thirty to forty actual men and their horses. Cavalry models should be arranged two to a base, with the base having a frontage of 40mm and a depth of 50mm.

Cavalry base and an overhead plan with dimensions.

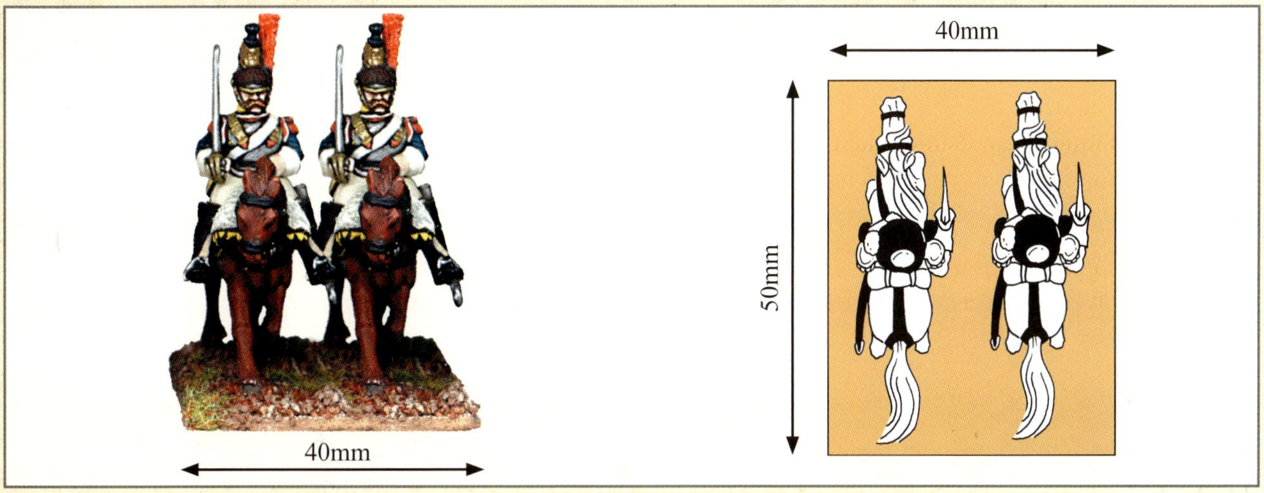

Three bases of Austrian Hussars.

Light cavalry units should consist of twelve models, while heavy cavalry units should consist of sixteen (possibly more for Guard units) but check the Army Lists for specific units. Include an officer, trumpeter and standard bearer model for an impressive effect.

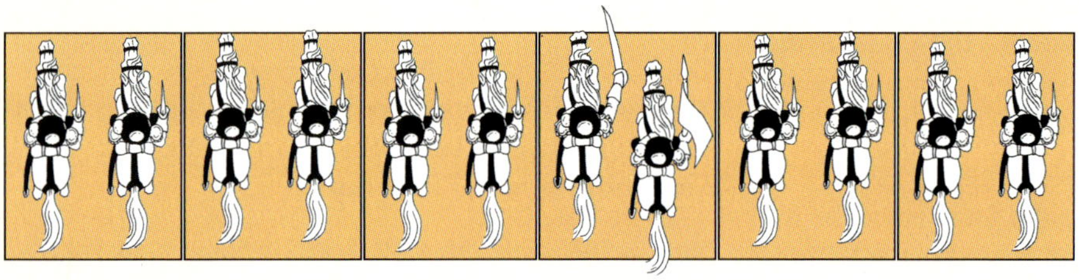

Light cavalry unit – 12 models on 6 bases – with command

French Light Cavalry Chasseur

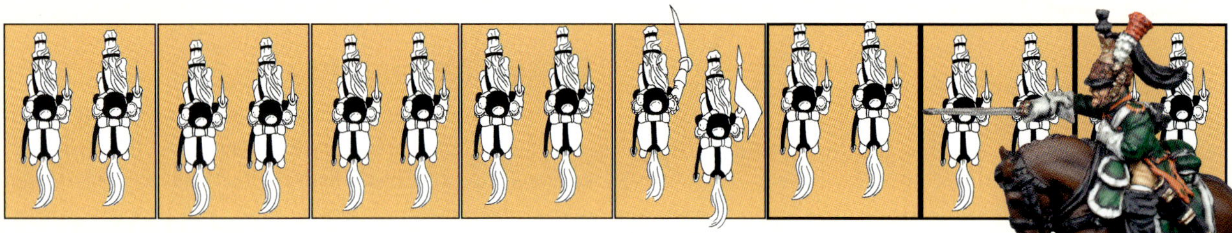

Heavy cavalry unit – 16 models on 8 bases – with command

French Heavy Cavalry Dragoon

Very large cavalry formations, such as the French Guard Chasseurs, will need to be fielded as multiple units. Therefore some Guard regiments can be treated as a whole division.

French Imperial Guard cavalry.

Artillery

Each cannon and crew model represents two actual guns and their complement of artillerymen; each cannon should be manned by three to six crew models, also mounted on the cannon base.

Four cannon models are enough for a typical unit, with each mounted individually on a base 50mm in frontage and 80mm in depth.

Artillery base and an overhead plan with dimensions.

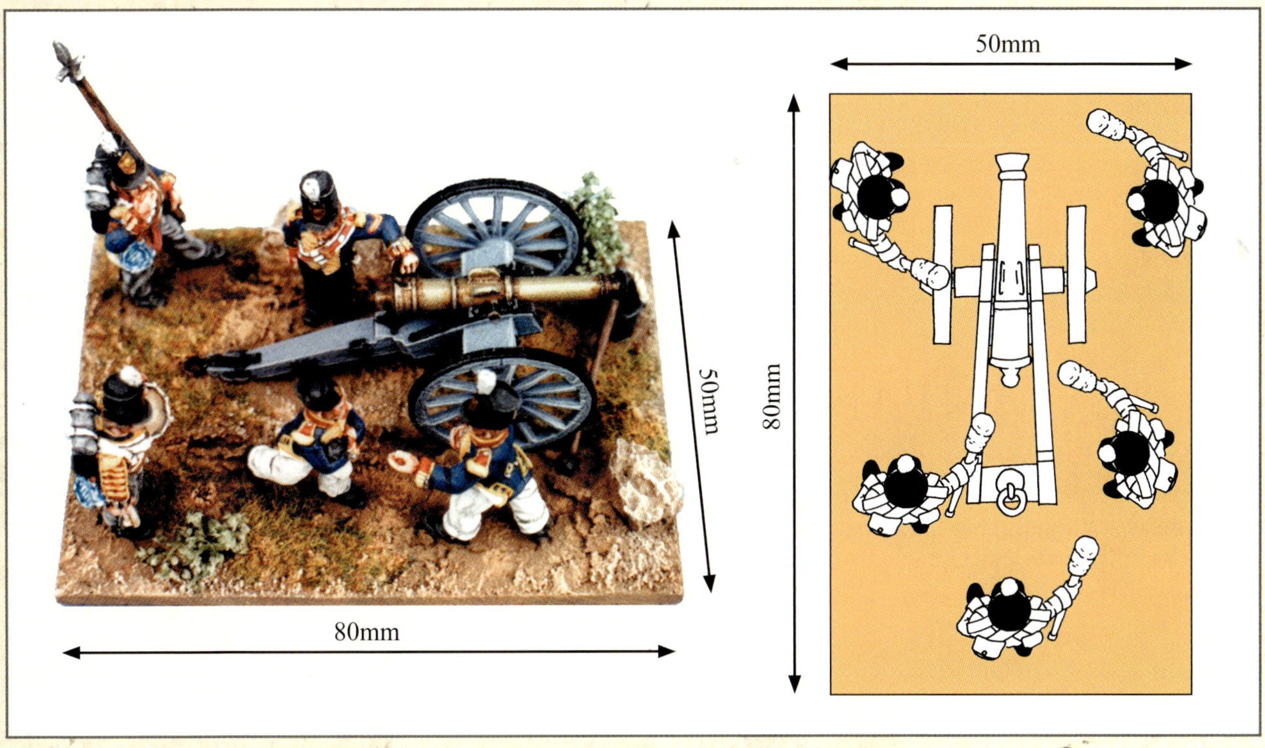

Additional crew can be added to the bases but they will have no effect on the game. A gun limber and horse team will also add flavour to the unit, as will the placement of mounted artillery officers, gabions and other assorted hardware, but these elements are purely cosmetic and are not required by the rules. It all depends on how much you enjoy painting and modelling.

> *God is on the side with the best artillery.*
> — Napoleon Bonaparte

Prussian Artillery consisting of four guns and crew.

FORMATIONS

It should not be believed that a march of three or four days in the wrong direction can be corrected by a countermarch. As a rule, this is to make two mistakes instead of one.

Prior to the Napoleonic Wars armies drew up in front of each other in huge lines, often with men deployed in ranks three or four deep. All this changed during the Revolutionary Wars when the French army took on masses of raw, poorly trained conscripts who found drilling and tactical discipline difficult to grasp. At the same time the size of armies across the whole of Europe was increasing to massive proportions. Generals found that their vast numbers of frequently ill-trained men could not recreate the large, drilled formations of previous eras and sought greater tactical simplicity.

During the Napoleonic Wars the number of ranks forming the depth of lines decreased (the British in particular deployed troops in ranks only two men deep) and new tactical formations were created to accommodate the new armies.

In keeping with the tactics of the period you must ensure that your units on the tabletop always conform to one of the following formations:

INFANTRY FORMATIONS

COLUMN OF MARCH

This is the fastest formation for manoeuvring units before they are required to deploy into fighting formations. The unit is deployed a single base wide. Although ideal for rapid movement, the unit may not fire in this formation and is very weak in close combat.

Infantry column of march.

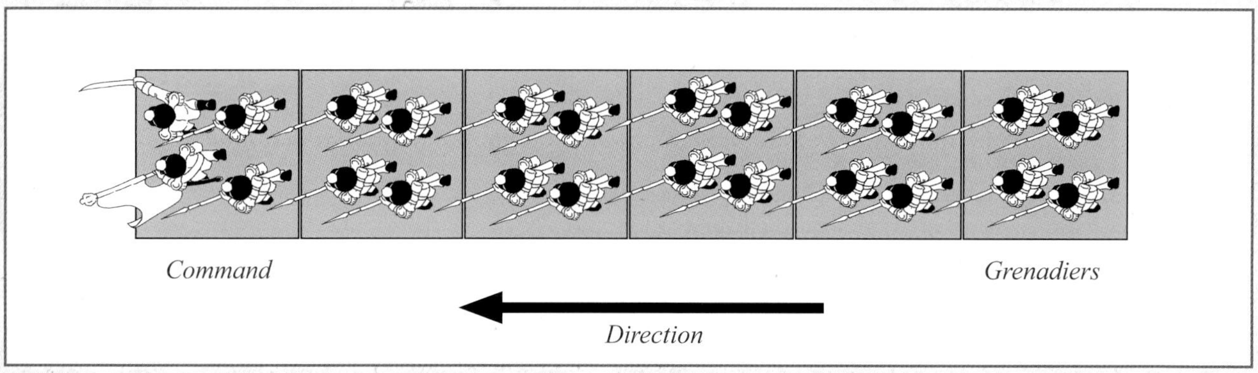

Infantry column of attack.

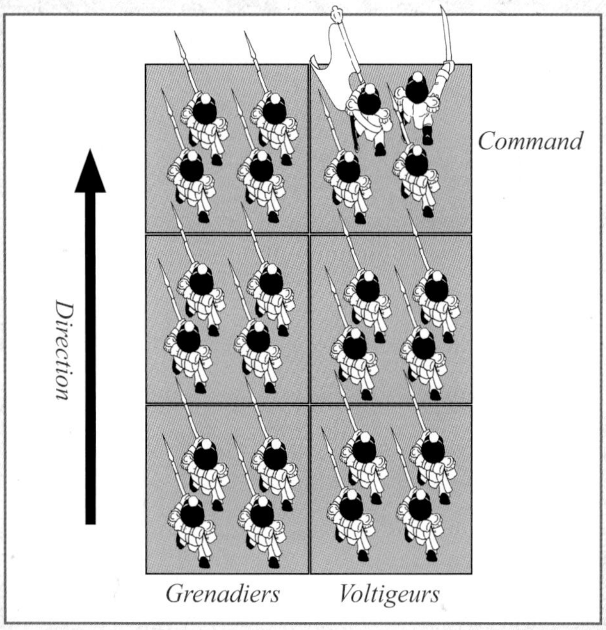

COLUMN OF ATTACK

The unit is deployed two or three bases wide, allowing for poorly trained troops to be almost herded around the battlefield. This formation has some fire power provided by the front of the unit, but its strength is in close combat. On impact the sheer number of men in such a tightly packed formation, with such depth, can crash through the enemy with devastating effect. This is the best formation for newly conscripted, poorly trained troops.

In war, the general alone can judge of certain arrangements. It depends on him alone to conquer difficulties by his own superior talents and resolution.

— Napoleon Bonaparte

LINE

This is the classic formation of earlier wars such as the Seven Years War. Line formation requires superior training and drill and takes up a lot of space on the battlefield. Troops are deployed either two or three men deep. The famous fighting formation of Wellington's army was two deep while more conservative armies tended to stick to the traditional three. This formation makes it difficult for the unit to move quickly, particularly if obstacles are in the way. Volley fire from lines can cause horrific casualties, often stopping charging units in their tracks. In game terms, a unit in line is deployed just one base deep.

Infantry unit in line.

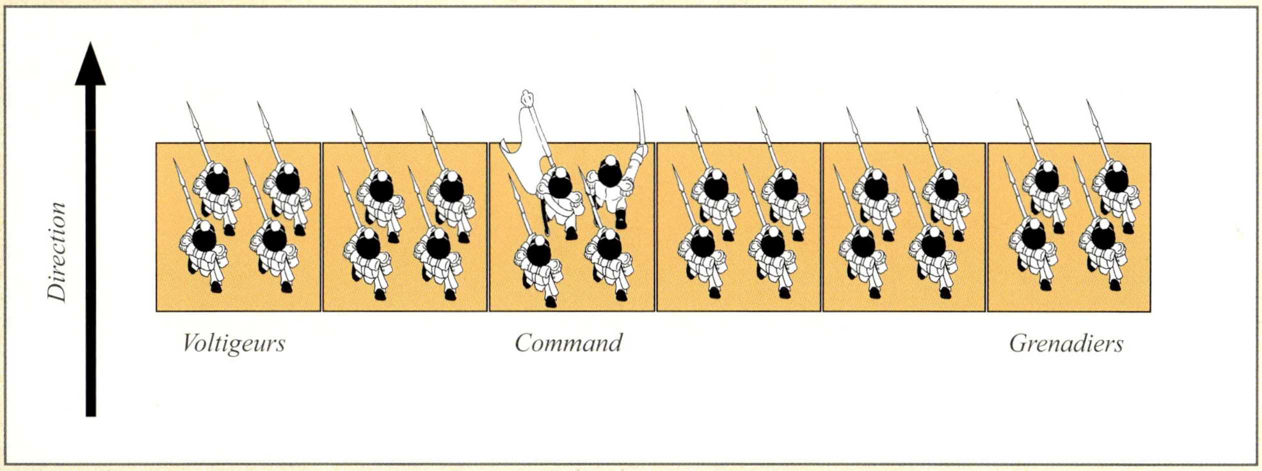

SQUARE

When facing armies containing thousands of cavalry, any infantry unit unfortunate enough not to be in square formation when charged by cavalry is almost certainly doomed. By forming a hollow square, with troops presenting bayonets and firing at the charging horses from all angles, the unit should be fairly safe. Squares were occasionally broken by cavalry, but it was uncommon. In game terms, square is pretty much as it sounds – arrange the bases so that all are facing outwards, with as near a 'square' shape to the unit as possible.

Unit in Square.

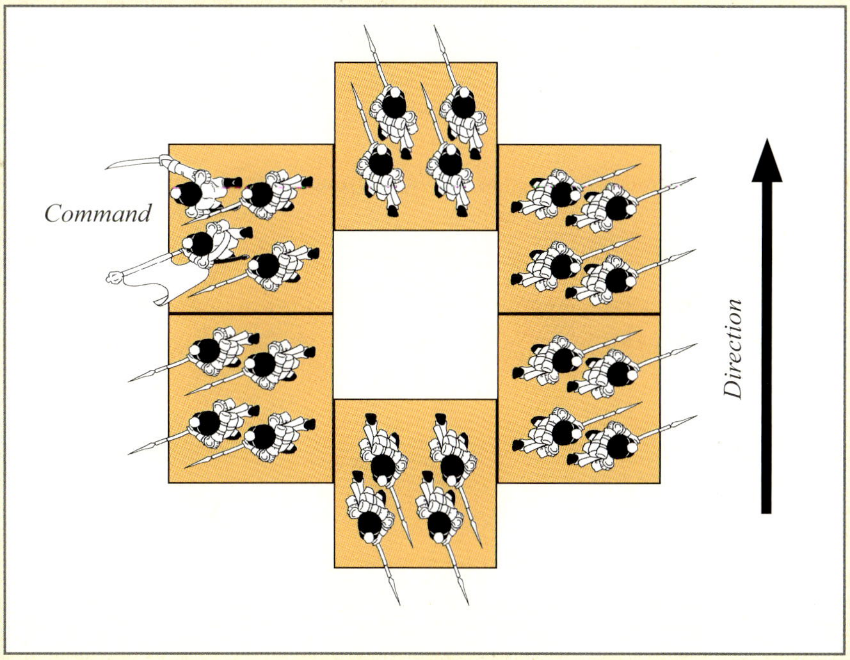

> *Charges of cavalry are equally useful at the beginning, the middle and the end of a battle. They should be made always, if possible, on the flanks of the infantry, especially when the latter is engaged in front.*
>
> — Napoleon Bonaparte

SKIRMISHERS

Many units had a company of light troops who were sent out in front, screening the main unit and providing fire power to break up enemy formations. The specific actions of light troops are not represented on the battlefield in the Basic Rules as they tend to clutter and complicate the tabletop. Full rules for light infantry and skirmishers can be found in the Advanced Rules section.

British Highlander.

Skirmishers detached from their parent light infantry (Chasseur) unit lead a mixed column.

Cavalry Formations

Cavalry Column of March

Cavalry can deploy in a single base Column of March in the same way as infantry. On good roads cavalry could cover vast distances in this formation, although good roads were a rarity in the early nineteenth century! The shorter distances present on the battlefield make this formation less useful for cavalry units and they are advised to immediately form one of the following formations upon arrival at the battlefield.

Cavalry unit in column of march.

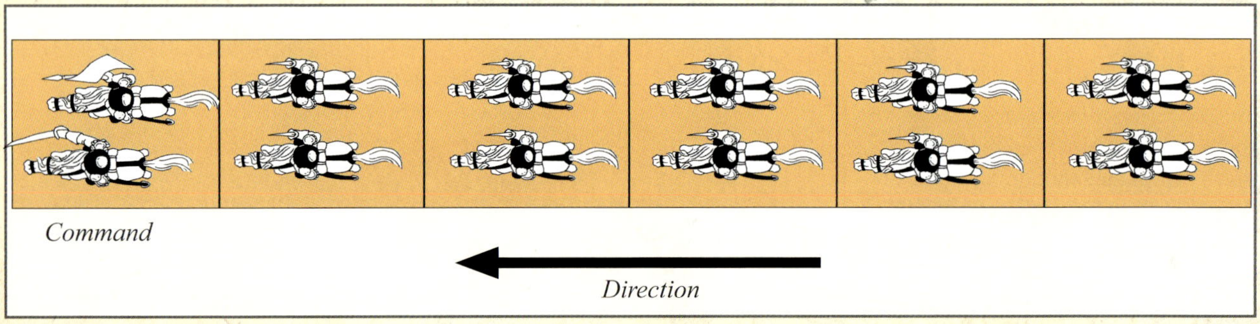

Cavalry unit in column of attack.

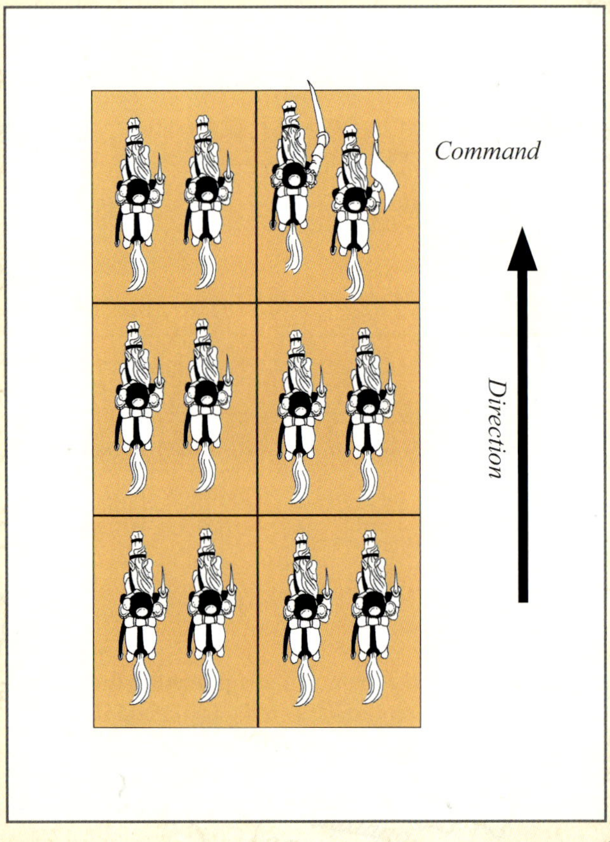

Cavalry Column of Attack

Riding two or three bases wide, this formation can be highly effective for heavy cavalry. The reaction to a tightly packed mass of big men, often wearing steel cuirasses, riding heavy horses, can be devastating to an infantry formation. French cavalry favoured attacking in this formation but be careful not to get over-confident though – cavalry columns are susceptible to heavy losses from artillery fire.

French Cuirassier in the rain.

Cavalry Line

The two horse deep line is the classic charging formation for cavalry. Deploying one unit next to another brings the maximum number of men into close combat. The momentum of men on horses obeying the final command to charge at a flat out gallop can devastate enemy infantry and cavalry alike. In game terms, this is represented by a cavalry unit lined up two bases deep.

Cavalry in line

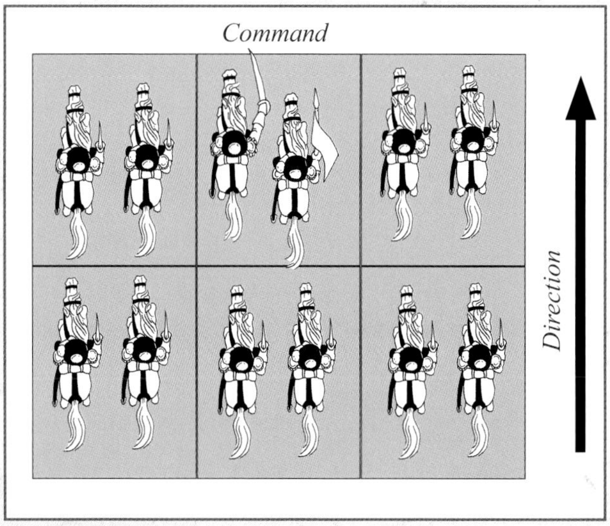

Artillery Formations

Artillery will always deploy in one of two formations (unless the crews have been forced to abandon their guns!); limbered or unlimbered.

Limbered

'Limbered' simply means in the process of being pulled by a team of horses; this was the most efficient way to drag the cumbersome artillery around the battlefield and although the guns move into position faster than if positioned by infantry, they are unable to fire while limbered.

Limbered artillery unit.

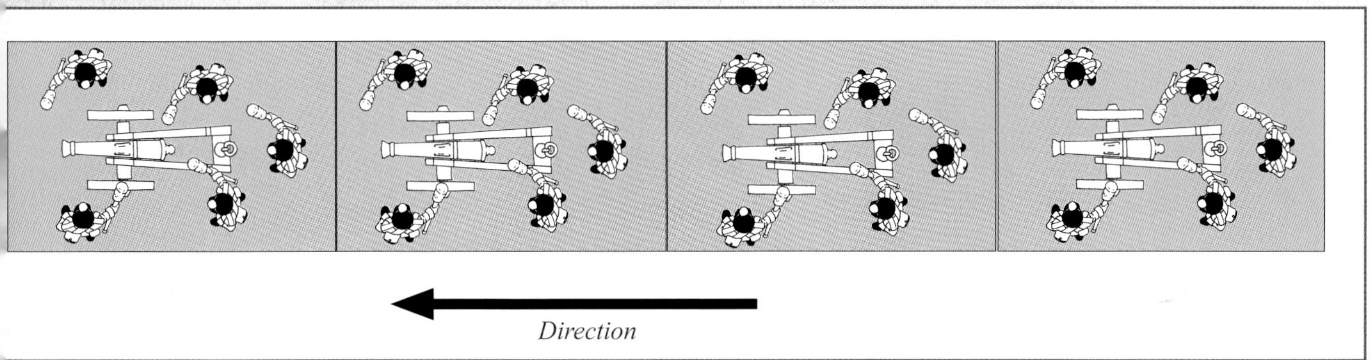

Unlimbered

'Unlimbered', conversely, means deployed, loaded and ready to fire. The guns can still be repositioned by hand, although this might require the help of some volunteers from a nearby infantry unit.

Unlimbered artillery unit 'Deployed Battery'

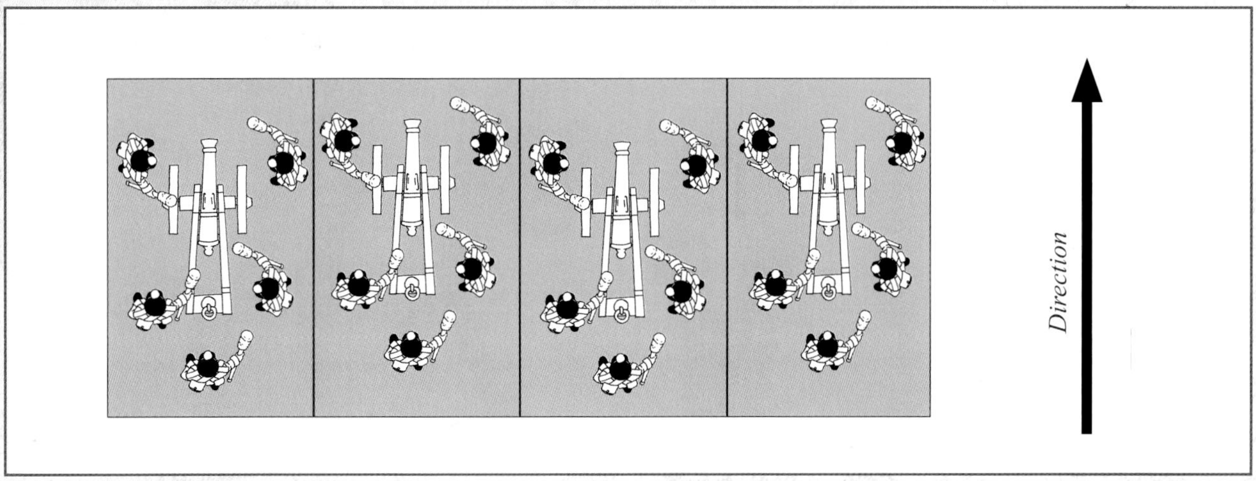

Artillery is more essential to cavalry than to infantry, because cavalry has no fire for its defence, but depends on the sabre.

— Napoleon Bonaparte

Prussian Landwehr Infantryman.

Deploying the Army

While the above formations apply to individual units, there is also the shape of the deployment of the entire army to consider. French armies were particularly adept at advancing in massive columns, closing upon the enemy in column of attack or forming lines to deliver devastating volleys of fire. Generals often deployed units to directly support each other. One cavalry unit, for instance, would be staggered slightly behind another, the front unit drawing enemy fire and leaving the rear unit (being relatively unscathed) to charge.

Another popular method of deployment was to arrange units in a chequerboard pattern. This formation had similar advantages to those above, but also left space for units to form square or deploy into line. Artillery and cavalry were reserved to give support to the main formation. Consider these choices when deploying your army for the coming battle, which we'll describe shortly.

French Chasseur a Cheval of the Guard (Knötel).

Before the Battle

If the art of war were nothing but the art of avoiding risks, glory would become the prey of mediocre minds. I have made all the calculations, fate will do the rest.

— Napoleon Bonaparte

Before each battle, certain preparations must be made – a battlefield must be chosen, and terrain placed upon it, scouts must conduct reconnaissance and armies must be deployed. This section is divided into three parts:

The Field of Battle

Rules for deciding what terrain will appear on the battlefield, and how it is placed.

Preparations

Determining the command rating of your generals, the percentage score value of your troops and the success of your scouts in reconnaissance.

Deployment

Placing your troops upon the field of battle, ready to begin!

The armies deployed ready for the start of the battle.

THE FIELD OF BATTLE

Soldiers generally win battles, generals get credit for them.

— Napoleon Bonaparte

Before battle can commence, you'll need to find a suitable playing surface to act as your battlefield. Ideally the gaming table should be 6' to 8' feet long and about 4' wide. Anything less than that will result in the opposing armies beginning the battle too close together, while a table over 5' wide can be a struggle to stretch across.

The easiest method of creating a gaming table is to lay a board of plywood, chipboard or MDF on top of your dining table. Paint the board grass green, or cover it with a sheet of green material, and you will have an instant battlefield. Most dining tables are a bit too small to act as the battlefield themselves, but they do provide an excellent and stable base. Never forget to place a sheet or some other kind of cushioned barrier between the wooden gaming board and your highly polished dining table. Failure to do so can result in battles of an altogether different kind!

To create a natural, realistic battlefield you will need to add some terrain and scenery in the form of hills, trees, hedges, walls, buildings, rivers and roads. 28mm model soldiers are only slightly larger than those used by model railway enthusiasts and a good model shop or website should be able to supply you with all manner of excellent, ready-made terrain pieces. Model railway manufacturers make some of the most amazing model buildings available. At first glance these might appear a little bit too small to accommodate your soldiers, but when spread out across the battlefield they look superb. You will have to trust your own judgement on the scale of terrain pieces, but bear in mind that this game uses 28mm models to represent men who, at this period in history, were on average about 5'6" tall. If all your terrain elements were exactly to scale with your soldiers then representing a Northern European pine forest would require model trees over half a metre in height!

Napoleon's Marshals discuss tactics.

Russian Pavlov Grenadier.

French Marshal with his troops outside his headquarters.

For those who enjoy the challenge of making their own scenic elements there are some quick and easy ways of producing effective terrain from readily available items. Twigs with kitchen scourers stuck on as leaves make excellent trees, especially with model railway flock sprinkled on to enhance the detail of the foliage. Roads can be made from sheets of sandpaper, or small pieces of gravel scattered on to the battlefield, while strips of cardboard painted blue and coated with several layers of gloss varnish make highly effective sections of river. If you are using a green cloth as the surface for your battlefield then you can place books or cut foam under it to create hills. Making scenery can become an absorbing aspect of the wargaming hobby; just remember not to get too carried away as you will need to leave plenty of space on the table to manoeuvre your armies.

French Hussars

Setting Up the Table

The players need to set up the battlefield in such a way that the layout of terrain is agreeable to everyone. Terrain affects the way your troops move and fight so it is important that players are not overly advantaged or disadvantaged by the scenery. If your game is part of an ongoing campaign run by a third party or referee then the referee will set up the battlefield and tell the players where to position their troops. Most games are fought as one-off battles though, and in these cases there are a couple of options to choose from:

1. One player sets up the scenery across the entire table and the other player chooses which of the long table edges he wishes to set up his army on. The player who sets the terrain then arrays his army along the opposite table edge.
2. All players agree to use the scenery generation system described below. This allows each player to take a hand in determining what the battlefield will look like – think of it as representing the armies manoeuvring into position.

Scenery Generation

Divide the battlefield into sections of 2 foot square. Each player should choose one long edge of the table on which to set up his army and then roll 1D6 for each section on his side of the table. For every roll of the die consult the chart below to see how many terrain pieces to set up in that section. If the table width results in an odd number of sections then the players should take it in turns to roll for the central sections of the battlefield. You might also wish to use the rules for Scouting Points on page 34 to determine the generation of scenery.

1D6 Score	Number of Terrain Pieces in the Section
1-2	No terrain.
3-4	One piece of terrain (roll on chart below).
5	Two pieces of terrain (roll on chart below).
6	The player may *choose* ***(no dice roll)*** one piece of terrain from the chart below.

Now you have determined which sections of the battlefield contain scenery you need to roll 2D6 for each terrain piece and check the chart below to decide the nature of the terrain. Where two pieces to are allowed (by rolling 5 on the chart above) terrain can be linked together or combined.

Example: A player rolls 5 on the above chart for one section of table, so he rolls twice on the following chart, scoring 4 and 5; a steep hill and a sparse wood. He decides to combine the terrain into a steep hill with a sparse wood on it.

Terrain can also be combined and interlinked between sections to form a continuous piece, if two adjacent sections are woodland for instance.

Russian units deployed in front of walled town, await the onslaught to begin.

Russian Officer.

2D6 Score	Terrain
2	A river or stream (player's choice) beginning at the player's table edge and running through one or more adjacent sections, avoiding hills and buildings. Roll 1D6 to determine how many sections in total the river runs through. If the score is 1 then the river does not cross any additional sections but enters and leaves the table in the same section.
3	An approximately circular area of dense woodland with a base diameter of between 6" and 12".
4	An approximately circular steep-sided hill with a base diameter of between 6" and 12".
5	An approximately circular sparse wood, vineyard or orchard with a base diameter of between 6" and 12".
6-7	An approximately circular gently sloping hill with a base diameter of between 6" and 12".
8-9	A 12" length of hedge, wall or fence arranged in a continuous, although not necessarily straight, line. This counts as a linear obstacle.
10	A single building with a ground area of approximately 6" x 6".
11	A farm consisting of between two and four buildings with a ground area of approximately 18" x 18", which counts as difficult ground.
12	An approximately circular area of bog, marsh, extremely dense scrub or woodland, or a steep hill with a base diameter of between 12" and 18", or a village with a minimum ground area of 18" x 18". All these count as extremely difficult ground.

Napoleon Bonaparte c.1800.

French Cuirassiers, surge forward from the edge of the woods, the British General has already deployed his infantry into the ultimate infantry defence against cavalry, the "square" (centre left).

There are certain things in war of which the commander alone comprehends the importance. Nothing but his superior firmness and ability can subdue and surmount all difficulties.
— Napoleon Bonaparte

PREPARATIONS

There is no man more pusillanimous than I when I am planning a campaign. I purposely exaggerate all the dangers and all the calamities that the circumstances make possible. I am in a thoroughly painful state of agitation. This does not keep me from looking quite serene in front of my entourage; I am like an unmarried girl labouring with child. Once I have made up my mind, everything is forgotten except what leads to success.

Before the game begins you will need to work out a few numbers for your army. These will have a direct effect on how your troops perform in battle and will help towards calculating the scale of victory or defeat in the aftermath:

1. Calculate the command rating of your Commander in Chief and any subordinate generals.
2. Work out the percentage score value of your infantry.
3. Determine scouting points.

1. Command Ratings

The command rating of a Commander in Chief or General represents how readily the troops under his command will accept his orders and act upon them, and how well he can motivate them in the heat of battle. The higher the rating, the more likely his troops are to respond positively to his orders and carry out his wishes in the face of the enemy.

There are three levels of commander: Commander in Chief, Divisional Generals and Brigade Commanders. All are collectively called 'generals' or 'commanders'; the highest ranking being the Commander in Chief. The Commander in Chief is in overall command of your entire force. He represents you on the tabletop and can be an army commander, corps commander or divisional commander, depending on the size of the battle you are fighting.

Below the Commander in Chief are a number of generals giving orders to their own divisions and brigades. The quality of commanders at this period of history varied greatly. Rank often depended on money, political position or membership of the aristocracy so the commander of an army was not always the most capable of its officers. It is more than possible in this game for lower ranking generals to have more talent for command than their senior officers.

The chart below gives the basic command ratings for the commanders of each nation during the Napoleonic period. You will notice that some change over time to reflect the changing fortunes of the war and the experience gained by the commanders in battle.

General's	**Command Ratings**					
Nationality	**1805-06**	**1807-08**	**1809**	**1812**	**1813-1814**	**1815**
French	10	10	9	9	8	8
Prussian	7	7	-	6	7	8
Russian	7	7	7	8	7	7
British	7	8	8	8	9	8
Austrian	7	6	7	6	7	7

Prussian Generals discuss the orders of the day.

You will now need to modify your generals' basic command ratings to create commanders that are truly unique.

Roll 2D6 for the Commander in Chief and choose the highest on either dice.

For Divisional Generals roll 1D6.

For Brigade Commanders roll 1D6 and deduct one from the score.

For each general check the die scores on the chart below and modify the basic command rating as directed.

Score	Command Rating Modification
1	Erratic General – roll again on this chart each game turn to give him a new rating.
2	Incompetent General – deduct 2 from his basic rating.
3	Poor General – deduct 1 from his basic rating.
4	Competent General – no modification to his basic rating.
5	Excellent General – add 1 to his basic rating.
6	Inspirational Leader – add 2 to his basic rating.

Example: A French General in 1812 has a basic Command Rating of 9. A Command Modification 1D6 roll gives a result of 3 – from the chart above he is classed as a Poor General and has 1 deducted from his Basic Command Rating thereby giving a final rating of 8.

His opponent is a Russian General who has a Basic Command Rating of 8. A Command Modification 1D6 roll gives a result of 6 – from the chart above he is classed as an Inspirational Leader and has 2 added to his Basic Command Rating thereby giving an improved final rating of 10 – more than a match for his French counterpart!

2. Percentage Score Value (PSV) of Infantry

During the Napoleonic Wars, the overwhelming number of infantry and their efficient use were seen to be the deciding factors in many battles. The Infantry formed the largest part of the army and, despite the heroics of the cavalry and the destructive firepower of the artillery, it was frequently the strength and deployment of the infantry that decided the outcome of engagements.

In the early stages of the Napoleonic Wars the infantry in the more organised armies were formed from professional drilled troops but as the conflict dragged on infantry numbers were increasingly made up of raw recruits and conscripts with little more than patriotic zeal to drive them. The proportion of infantry in most armies steadily grew (this was especially true for the French in later campaigns) and they were increasingly relied on to decide the outcome of the battle.

Before battle commences you need to work out the strength of your army. This will become important later on when deciding on the winner of the battle and in determining how decisive the victory is.

The strength of an army is based solely on the quantity of infantry troops it contains and to calculate the PSV simply divide 100 by the number of infantry units in your army – rounding any fractions up or down to the nearest whole number.

Example: An army contains nine infantry units; dividing 100 by nine gives to each infantry unit a PSV of 11.11, rounded down to 11.

3. Scouting Points

Finally, you have to determine which army has carried out the superior reconnaissance on its enemy's disposition. Most of this task was performed by an army's cavalry patrols, especially the lighter and more nimble mounted troops such as Light Dragoons, Chasseurs, Chevaux-Legers and Hussars.

Each player should add up his army's scouting potential:

- Each light cavalry unit in the army equals two points.
- Each heavy cavalry unit equals one point.

Then roll a 1D6 and add the army's total scouting points to the score. This will give you your total scouting points.

Example: An army has three light cavalry units and one heavy cavalry, giving seven scouting points in total. The result of the 1D6 roll is 3, giving a combined total of 10 scouting points.

If you wish you can use scouting points to help determine the placing of scenery on the tabletop – the player with the most scouting points being the one to decide where scenery is placed and/or choosing which table edge he wants to deploy his army along once the scenery has been laid out.

This would be logical, as it would assume that the side with the higher scouting points would have a greater knowledge of the battlefield landscape but this is purely at the players' discretion and not an obligatory part of the rules.

I believe one bad general to be worth two good ones.

— Napoleon Bonaparte

British Infantryman.

DECIDING THE OVERALL WINNER

Victory belongs to the most persevering.

— Napoleon Bonaparte

At the end of every turn each player should work out their army's current victory score.

When a player has captured or routed 60% or more of his opponents army the game is over; it is then necessary to calculate the number of Victory Points for each player to determine the winner and the extent of the victory.

As well as the PSV Victory points additional points are earned for the capture of enemy Regimental Colours, Eagles or cannon (taking such trophies and displaying them in their cities was an ideal way for generals to boost their prestige, and their political ambitions).

You can capture enemy infantry and heavy cavalry colours by routing the unit from close combat; this occurs when a unit on lost command orders loses in close combat. Cavalry units who defeat infantry troops not formed in square in close combat automatically capture the infantry colours.

Your army receives 5 victory points for each enemy unit's colours they have captured.

Artillery batteries that are beaten in close combat lose their guns; for each artillery battery your army captures you receive 5 points..

Points Difference	Extent of Victory
40 points or more	Decisive victory.
30-39 points	Good victory.
20-29 points	Victory.
10-19 Points	Winning draw.
0-9 points	Draw.

Captured enemy colours and gun.

Prussian Generalfeldmarschall Gebhard Leberecht von Blücher, after much perseverance celebrates with his high command the final victory over Bonaparte.

French Officer.

DEPLOYING YOUR ARMY

Never lose sight of this maxim, that you should establish your cantonments at the most distant and best protected point from the enemy, especially where a surprise is possible. By this means you will have time to unite all your forces before he can attack you.

— Napoleon Bonaparte

When deploying your army on the tabletop you, and your opponent, must decide which of your divisions or brigades to deploy first, and then establish the order in which to deploy the rest of your army.

The player with the greatest number of scouting points can decide whether to deploy his initial division first or make his opponent deploy first.

Players then alternate putting their divisions on the battlefield one at a time; this allows players to make strategic and tactical decisions about the deployment of their army based upon the troop positions of their opponents.

When playing your first few battles, deploy your entire army on the table before the battle starts. Rules for reinforcements and flank attacks can be found in the Advanced Rules section which allow some units to arrive after the battle has commenced.

RULES OF DEPLOYMENT

When deploying your army you must place all your units along your own side of the table edge with no models within twelve inches of the table's centre (unless specifically allowed by a scenario, campaign or a historical re-fight). Your units can be deployed in any of the unit formations previously mentioned. Artillery batteries can be deployed (they do not have to start the game limbered) and skirmishers may be detached (if you are including the Skirmisher Rules in your game).

CASUALTY REMOVAL

In these rules, models representing casualties are never removed from the battlefield until a whole unit is destroyed. After spending hours painting your armies you want to keep them on the table for as long as possible!

A unit is usually destroyed after a second morale test is failed (third for Guards) at which point all the models in the unit are removed together.

Casualties can be recorded on the unit by placing special casualty models, rocks or spare dice behind the appropriate unit.

A general-in-chief should ask himself several times in the day, 'What if the enemy were to appear now in my front, or on my right, or my left?

— Napoleon Bonaparte

Casualty markers, modelled bases of casualties.

Casualty markers, dice behind the unit.

Russian Drummer.

It should not be believed that a march of three or four days in the wrong direction can be corrected by a counter march. As a rule, this is to make two mistakes instead of one.

— Napoleon Bonaparte

French Marshal (Knötel).

The Battle

When you determine to risk a battle, reserve to yourself every possible chance of success, more particularly if you have to deal with an adversary of superior talent, for if you are beaten, even in the midst of your magazines and your communications, woe to the vanquished!

— Napoleon Bonaparte

In this section we cover all the things you will need to know to actually fight out the battle. Unlike the previous section, which covered a lot of basic first steps, you will use the information presented here every time you fight a battle. Over time you'll become familiar enough with the rules to play from memory, but there will still be occasions when you'll quickly need to look something up. For that reason, we've presented the rules in the order in which they are used

French Dragoon (Knötel).

Bavarian Infantryman.

TURN SEQUENCE

My rule always was to do the business of the day in the day.
— The Duke of Wellington

The game play is divided into **turns**, during which both armies move, fire their weapons and engage in hand-to-hand combat. In order to do this each turn is sub-divided into **phases** that have to be carried out in a specific order.

Once both players have been through all the phases separately the turn is complete and a new turn begins.

The sequence of phases is as follows:

1. **Roll for Initiative.**
2. **Place Command Cards**
3. **Artillery Fire**
4. **Reveal Command Cards**
5. **Movement**
6. **Skirmisher Fire Resolution**
7. **Volley Fire Resolution**
8. **Close Combat Resolution**
9. **Morale Checks**
10. **Determining the Winner**

In the Spirit of the Game

These rules try to cover all the situations and events that might occur during the course of a game, but occasionally a situation may arise for which the rules do not provide a definitive answer. Players should normally amicably decide and agree between them how the problem should be resolved; if this proves impossible then each player should roll a 1D6 and the highest scoring player can decide. More important than anything else is that games are at all times played in the proper spirit – one of fair competition and friendly rivalry. Disputes over the rules are not the way to prove superior generalship.

1. ROLL FOR INITIATIVE

In war there is but one favourable moment; the great art is to seize it!
— Napoleon Bonaparte

The player with the higher initiative score chooses whether to go first or second. To determine initiative score each player rolls 1D6 and the highest number wins the initiative for the turn (in the event of a tie, continue to roll until a winner is found).

You must roll again for initiative to determine who goes first or second **at the start of each new turn sequence.**

French Guard Chasseurs a Pied, the light infantry of the Imperial Guard.

2. Place Command Cards

Strategy is the art of making use of time and space. I am less concerned about the later than the former. Space we can recover, lost time never.

— Napoleon Bonaparte

Command cards represent tactical orders – orders given to units when they are very close to the enemy and ready to attack, or in danger of immediate attack themselves. At other times when manoeuvring more distantly, units won't require command cards at all. You can think of a command card as representing the last hasty order a general is able to give before his men engage the enemy.

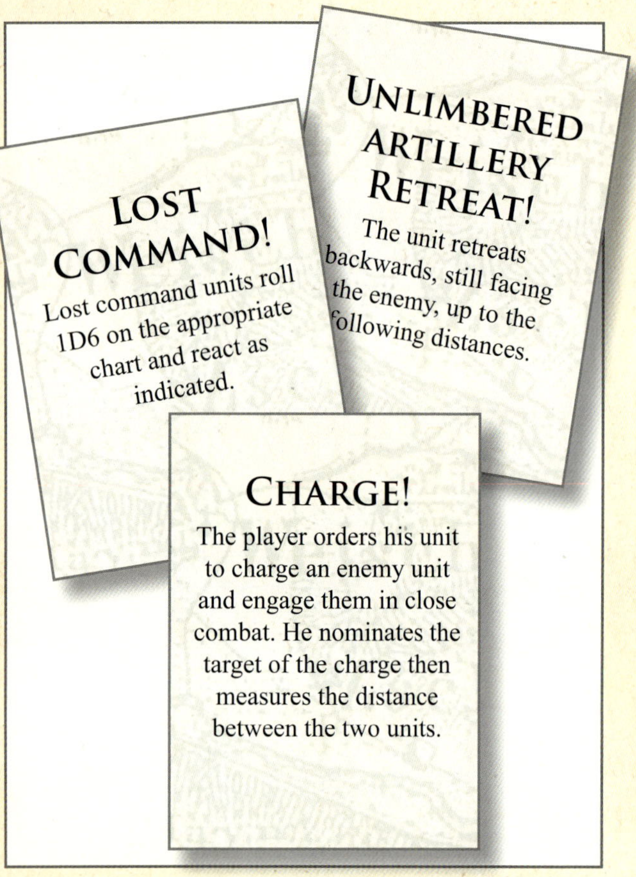

Engagement Range

Command cards only come into play once opposing units start to get to grips with each other; that is, once they enter into **engagement range**. As the army Commander in Chief you will be faced with the task of giving orders to your units who are about to enter into close combat with the enemy. The action at this point becomes fast-paced and you will have to make on-the-spot tactical decisions to get the best out of your troops. At this point you will see if you have manoeuvred them into position skilfully!

The term engagement range refers to the area on the battlefield in which a unit is likely to come into contact with the enemy. The engagement range is double the unit's standard movement rate. Movement rates are discussed in full in the next phase but are summarised here along with engagement ranges:

Unit	Standard Movement Rate	Engagement Range
Light cavalry	8"	16"
Heavy cavalry	6"	12"
Infantry	4"	8"
Artillery	4"	8"

All units able to charge or be charged are considered to be within engagement range. On occasion, this will effectively increase a unit's own engagement range if they themselves are within range of a faster unit.

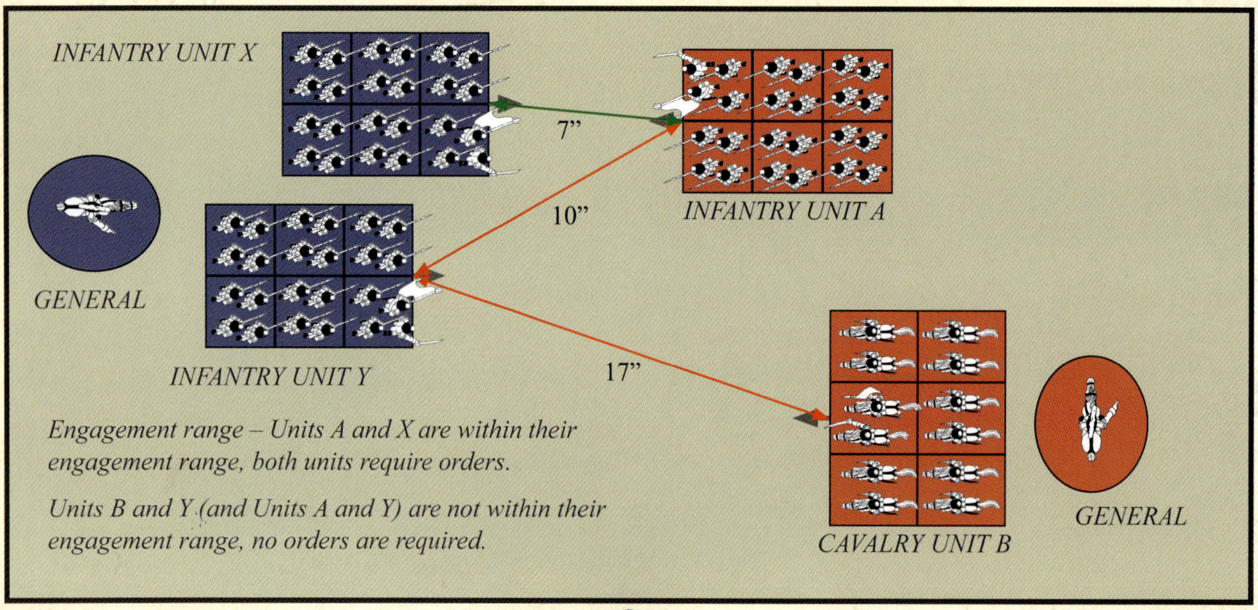

Engagement range – Units A and X are within their engagement range, both units require orders.

Units B and Y (and Units A and Y) are not within their engagement range, no orders are required.

Orders

All units that are within engagement range must be given orders at the beginning of the turn sequence, before resolving artillery fire. This is done by giving all the units within engagement range a command card. You should choose the order that you want to issue to the unit and place the appropriate command card face down behind the unit without your opponent seeing what the order is. A sheet of command cards to use in your games is provided with these rules for you to photocopy or cut out, and you can also download them from the Foundry website: **www.wargamesfoundry.com.**

Napoleon and Marshal.

Command Cards

The following command cards are available to Divisional or Brigade Generals:

Charge!

Run!

Retreat and Fire!

Stand and Fire!

Retreat!

Advance and Fire!

Form Line!

Form Square!

Form Column!

Unlimbered artillery has fewer command options available. They can

Stand and Fire!

Run!

Retreat!

Limbered artillery can only Retreat! or Run!

The effects of each of these orders is summarised on the cards and described in full later.

3. Artillery Fire Phase

The worse the troops the greater the need of artillery.

— Napoleon Bonaparte

Both players take it in turn to fire their artillery batteries during this phase, the player with the Initiative firing one of his batteries first, and then his opponent returning fire with one of his batteries. Both armies continue to fire alternately in this fashion until all artillery batteries have fired.

The armies of the Napoleonic wars employed a plethora of different artillery guns. For the sake of simplicity we divided artillery into three main categories: **heavy artillery**, **field artillery** and **horse artillery**. In addition to these conventional artillery pieces, there is also the **howitzer**, a special type of gun somewhat akin to a cross between a cannon and a mortar; howitzers had a very short barrel and could launch their shells high into the air.

Of course, the variety of individual artillery pieces available is huge and categorising them can sometimes be difficult. If you're new to Napoleonic wargaming, you might like to take a quick look there to familiarise yourself with some of the many varied artillery types (and their miniature equivalents!) but for now all that's really important is to remember that all artillery pieces, no matter how unusual, fall into one of these four categories: heavy artillery, field artillery, horse artillery or howitzer.

Ranges

The different categories of cannon have the following firing ranges:

	Close Canister	Canister	Effective Range	Long Range	Extreme Range
Heavy Artillery	0 to 6"	6+ to 12"	0 to 24"	24+ to 36"	36+ to 60"
Field Artillery	0 to 6"	6+ to 12"	0 to 24"	24+ to 36"	36+ to 48"
Horse Artillery	0 to 6"	6+ to 12"	0 to 24"	24+ to 36"	36+ to 48"

Howitzers have the same ranges as field artillery, though fire differently (see Howitzers, page 44).

CHOOSING A TARGET

Declare your intention to fire upon a specific unit. Artillery have two means of firing: shot (which are large, heavy cannonballs) or canister (which unleashes a 'grapeshot' effect of many dozens of smaller musket balls hitting at close range). Choose which type of ammunition you are firing when declaring your target.

Firing

When firing howitzer or shot you can choose any enemy unit as a target provided it is within range, within arc of fire and in direct line of sight of the firing battery.

When firing canister the target is always the nearest unit in direct line of sight and the range of fire is measured independently from the centre of each individual base. The arc of fire is the area within two lines drawn at 90 degrees from the centre of each individual base.

A battery cannot divide its fire between different targets (unless firing canister) and cannot divide its type of fire between shot and canister except in the case of howitzers, which fire different types of ammunition.

Range

Range measurements are taken from the centre of the artillery battery to the centre of the unit it is firing upon.

Arc of Fire

Arc of fire for artillery firing shot is the area within two lines drawn at an angle of 45 degrees from the front end corner of each battery (see diagram).

Arc of fire for artillery firing canister is the area within two lines drawn at a 90 degree angle from the centre of **each individual artillery base**.

Line of Sight

Line of sight is literally what you can see from the artillery's position – get down over the table and have a look! Use common sense: give the benefit of the doubt to artillery

Choosing a Target – Howitzer and Shot.

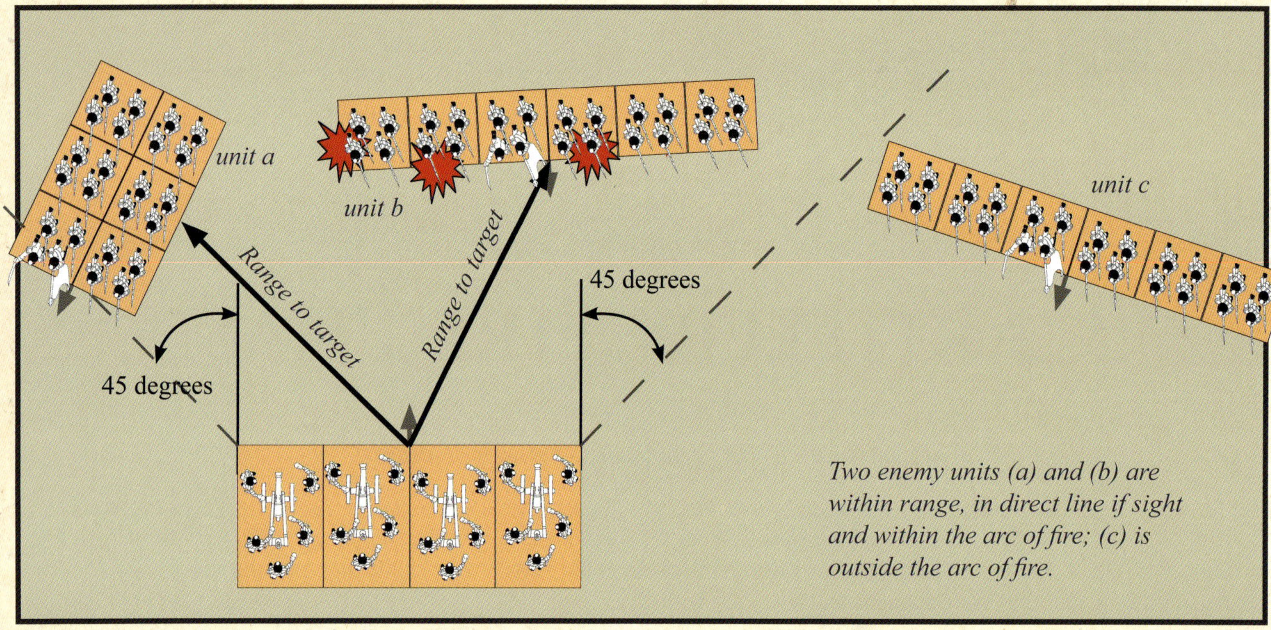

Choosing a Target – Canister Range.

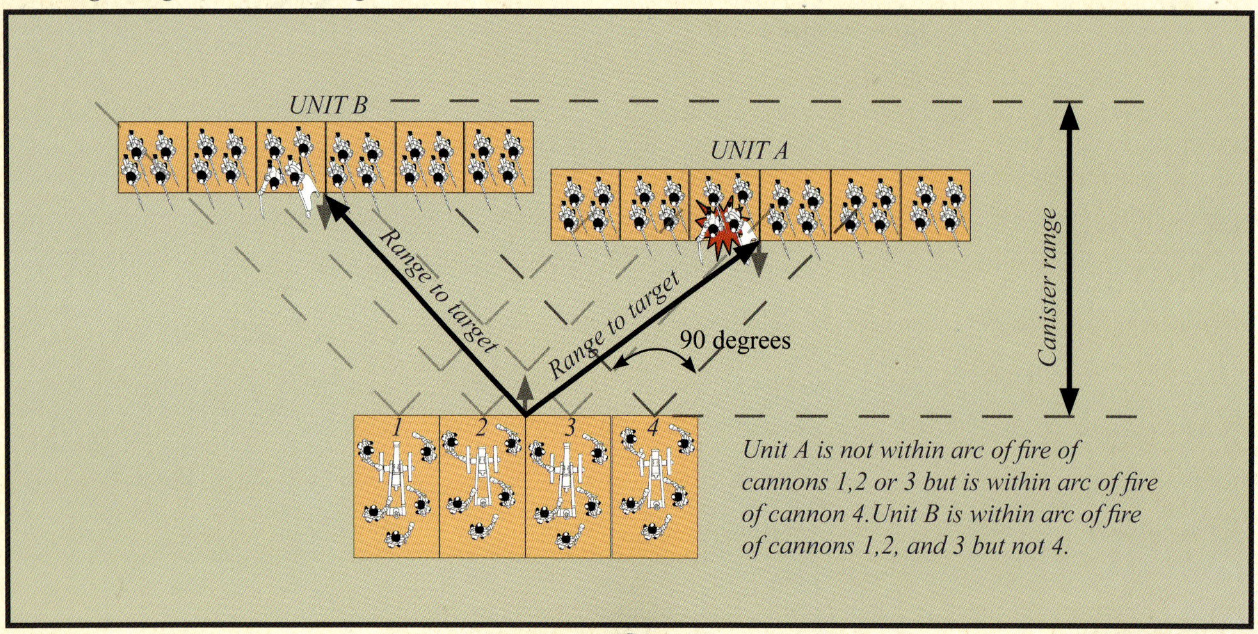

positioned high up on hills; conversely, units shouldn't be allowed to fire just because they can see the points of bayonets or the tops of hats!

Exceptions

You cannot fire:

1. Into an existing close combat situation.
2. Over your own troops unless the firing artillery is deployed on the front half of a hill. Howitzers can fire shell (but not canister) over friendly troops no matter where they are deployed if targeting an enemy at long or extreme range.
3. Through woods (you can fire up to 2" inside the edge of a wood).
4. At troops deployed over the crest of hills (over half way back from the front of the hill).
5. At troops deployed more than 2" behind hedges, walls, fences or any other intervening linear obstacles unless your guns are positioned on the front half of a hill higher than the obstacle.
6. Over houses or villages.

Artillery Firing Procedure

Using the following procedure when firing solid ball shot – i.e., when firing at effective, long or extreme range. Don't use this procedure for canister fire, which is described later.

Determine the target unit's formation (the frontage of the unit being targeted will determine how difficult it is to hit) and whether it is being fired upon in the front, rear or flank.

There are two charts below – one for target units hit from the front or rear, and one for target units hit in the flank. Find the correct one and roll 1D6 for each gun model in the firing battery. Check the score against the appropriate range and type of target unit. The number shown is **the minimum dice score needed** for the gun to hit its target; if the score is less than that required then the gun has missed.

Austrian Cannon.

Firing into the Front or Rear

	Score Needed to Hit		
Target Unit	**Effective Range**	**Long Range**	**Extreme Range**
Artillery Battery	6	6*	-
Rifle Detachment	6	6	-
Cavalry Line	6	6	6*
Cavalry Column	5+	6	6
Infantry Column	5+	6	6
Infantry Square	4+	5+	6
Infantry Line	4+	5+	6*

Firing into the Flank

	Score Needed to Hit		
Target Unit	**Effective Range**	**Long Range**	**Extreme Range**
Artillery Battery	6	6*	-
Rifle Detachment	6	6*	-
Cavalry Line	6	6*	-
Cavalry Column	5+	6	6*
Infantry Column	5+	6	6*
Infantry Square	4+	5+	6*
Infantry Line	6	6*	-

* Against targets marked with an asterisk (*) heavy artillery inflicts only a single casualty per hit, rather than the normal two. This is explained below, in Effects of Artillery Fire.

British Infantryman.

EFFECTS OF ARTILLERY FIRE

Every hit from artillery fire causes a casualty on the target unit. In the case of heavy artillery, each hit causes two casualties on the target unit, except against targets marked with an asterisk (*) on the above tables where only a single casualty is caused.

The additional damage from heavy artillery is down to two factors: the size of gun and the formation of the unit under fire. The heavier the weight of the cannon balls, the further the ball will penetrate into the ranks, and the deeper the formation the worse this effect will prove to be, hence the greater casualties suffered by units in column.

In summary:

Heavy Artillery	2 casualties per hit (1 casualty per hit where marked *).
Field Artillery	1 casualty per hit.
Horse Artillery	1 casualty per hit.

Place a marker on the target unit to show the number of casualties. Don't remove individual bases to represent the casualties; the effect of the losses suffered by the unit will be reflected in its morale as described later.

Hits on Enemy Artillery

When firing at an enemy artillery battery within the effective range there is a chance that you might actually destroy a gun instead of causing casualties amongst the crew. Roll a further 1D6 for each successful hit on an artillery battery; on a score of 6 the gun is destroyed and may take no further part in the battle. The crews may continue to fight with any other guns in the same battery.

HOWITZER FIRE

Howitzers do not fire shot, but instead fire shells. When firing a shell each howitzer model can select any unit it is permitted to fire upon (due to range, arc of fire, line of sight, etc.) as a target. When firing canister, howitzers must fire at the same target as the rest of the battery. Ranges for howitzers are the same as for field artillery with the distance measured from the centre of the howitzer model to centre of the target unit.

Prussian Stragglers!

To determine whether a hit is scored, follow the same Firing Procedure as for other guns, except 2D6 are rolled per gun instead of one. The score required to hit is the same as for field artillery. Roll 2D6 for each howitzer.

1. If two hits are scored, then a direct hit and explosion has occurred and the target unit suffers 1D6 hits. The firer then rolls a number of dice equal to the number of hits and any dice which scores a 5 or 6 causes a casualty. Troops deployed inside buildings, farms or villages receive a save from being in cover, but all other saves are ignored.
2. If one hit is scored, a glancing blow has occurred and the target unit suffers 1D6 hits. The firer rolls a number of dice equal to the number of hits and any dice which scores a 6 causes a casualty. Troops deployed inside buildings, farms or villages receive a save from being in cover, but all other saves are ignored.
3. If either dice scores a 1, the fuse fails to light and no hits are inflicted, regardless of the score on the other dice.
4. If both dice score 1, the lit fuse has prematurely exploded inflicting a single casualty on the firing artillery battery.

CANISTER (CASE/ GRAPESHOT) FIRE

At close range, gunners of the Napoleonic period switched from firing their solid ball shot or shell and employed a devastating round of canister or grape shot.

When firing canister the battery targets the nearest unit to each gun model, so in this case a battery can effectively divide its fire between different targets. Target units suffer 1D6 of hits from each gun firing upon it. The firer then rolls a number of dice equal to the number of hits and requires a score equal to or better than shown on the chart below to cause a casualty:

	Close Canister	Canister
Heavy Artillery	3+	4+
Field Artillery	4+	5+
Horse Artillery	4+	5+

Troops deployed inside buildings, farms or villages still receive the save from being in cover, but all other saves are ignored. When firing canister, the target's unit type and formation has no effect on the chances of being hit – grapeshot launches such a horrific mass of shot that it simply can't be avoided.

ARTILLERY MORALE EFFECT

The horrific injuries and number of casualties caused by artillery fire could break the resolve of many troops, so any units taking a hit from canister or two or more hits from shot and shell (howitzer) fire must take an immediate morale test. Morale tests are described on page 58.

4. Resolve Command Cards

Never interrupt your enemy when he is making a mistake.

— Napoleon Bonaparte

Reveal Cards

Once artillery fire has been resolved the players reveal their command cards. The player with the initiative for this turn chooses one of his command cards to reveal, and then rolls to see if the order is successfully carried out.

Test to Give Out Orders

When a unit's command card is revealed it means only that the general has issued the order to the unit in question, it does not necessarily mean it will be received and obeyed. You must therefore test to see if the troops respond to the order.

Each unit with an order should roll the number of dice shown below and compare the score to their General's command rating. Units use their Divisional or Brigade General (referred to simply as 'their General') and roll the number of dice as follows:

The unit nearest to the divisional or brigade general rolls 1D6.

Units within 8" (infantry) or 16" (cavalry) of their divisional or brigade general roll 2D6.

Units more than 8" (infantry) or 16" (cavalry) away from their divisional or brigade general roll 3D6.

Add up the scores on the dice and apply the following modifiers if applicable:

For every unit that has been destroyed or become detached or is running or routing from a division, the remaining units in that division must add 2 to their total dice score when testing to receive orders.

Divisions with a starting strength of five or more units can ignore the first unit that is destroyed, detached, running or routing. This rule does not apply to Guard or elite divisions.

Results

If the total score is **lower** than the command rating of the unit's General, the unit receives the order successfully and attempts to carry it out.

If the dice result is **equal** to or **greater** than the command rating of the unit's General then the unit has failed to receive or understand the order and must remain stationary for the rest of the turn.

Now you can see why a Division or Brigade General must stay close to the units in his division when they are caught up in the thick of battle!

Further Orders

If the test to give out orders is successful, the unit will immediately carry out its order as instructed. The effects of these orders are described on page 46-48. If the test is unsuccessful, the unit will remain stationary. Either way, play now passes to the other player who may now reveal one of his own command cards.

Reactions

When an order is issued, enemy units nearby sometimes have the option of reacting. The exact way in which an enemy unit can react is detailed in the description for each individual order (below). If a unit does react, that reaction will count as the player's turn and that same player cannot then reveal another of his command cards until play has first returned to the first player. In short, each player has the opportunity to either react, or to reveal one of his own command cards as normal. He cannot do both.

An extraordinary affair. I gave them their orders and they wanted to stay and discuss them.

— Duke of Wellington

British Highlander Command.

COMMAND CARD ORDERS

The secret of war lies in the communications.

— Napoleon Bonaparte

When a command card order is successfully issued to a unit, its effects are resolved as follows:

CHARGE!

The player orders his unit to charge an enemy unit and engage them in close combat. He nominates the target of the charge then measures the distance between the two units. If the target unit is within Standard Movement rate, the player moves the charging unit into base-to-base contact with the enemy. The units are thereafter deemed to be in Close Combat, the outcome of which will be determined in the Close Combat Phase.

The defending unit has its command card removed, as their attackers have fallen on them before they can react.

Charging Distances:

Unit	Standard Movement Rate	Charging Distance
Light cavalry	8"	16"
Heavy cavalry	6"	12"
Infantry	4"	8"

Counter Charging.

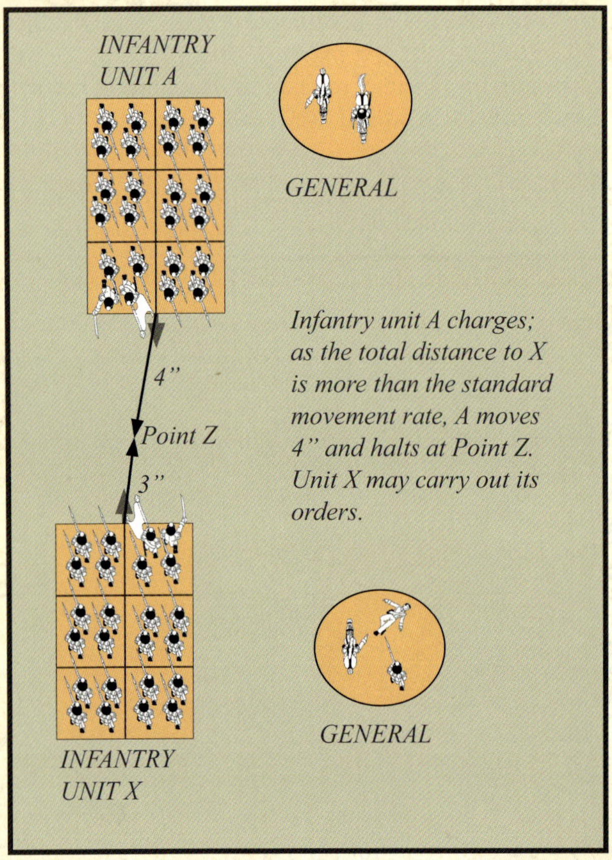

Reacting to Charges

If the charging unit moves more than it's standard movement rate (i.e. more than half of its charge distance) the target has the opportunity to react and may play its command card.

If defending unit chooses to reveal its command card, the order is carried out as normal after which the unit can do nothing more in this particular phase and play returns to the player who launched the initial charge, who may now choose another unit to reveal a command card for.

A unit being charged is not obliged to react, even if able to do so. The unit may choose simply to accept the charge and not reveal its command card until later.

Firing at Charges

Any charging units sustaining two or more casualties due to enemy Stand and Fire should take an immediate morale test in order to press home the charge (inflicting casualties in a desperate volley at the last possible moment may repel the charge).

- If the charging unit passes the morale test the unit presses home the charge and enters into base-to-base contact with the enemy.
- If the charging unit fails the morale test it fails the charge and should be placed 1" away from the defenders.

Failed Charges

Failed charges can occur for a variety of reasons:

- If the charging unit chooses a target beyond its charge distance.
- If the charging unit has failed a morale test resulting from an enemy unit's Stand and Fire! reaction.
- If the defending unit it is attempting to charge has moved further away as the result of a reaction.
- When terrain unexpectedly reduces the movement rate of the charging unit.

If a charging unit is too far away to complete the charge, move it the full charge distance towards the defending unit and then halt. If a charge fails for any other reason, the charging unit halts 1" away from their target.

A unit that makes a failed charge automatically rolls a 1D6 and suffers casualties as follows:

- Scores of 1 or 2 equals one casualty.
- Scores of 3 or 4 equal two casualties.
- Scores of 5 or 6 equal three casualties.

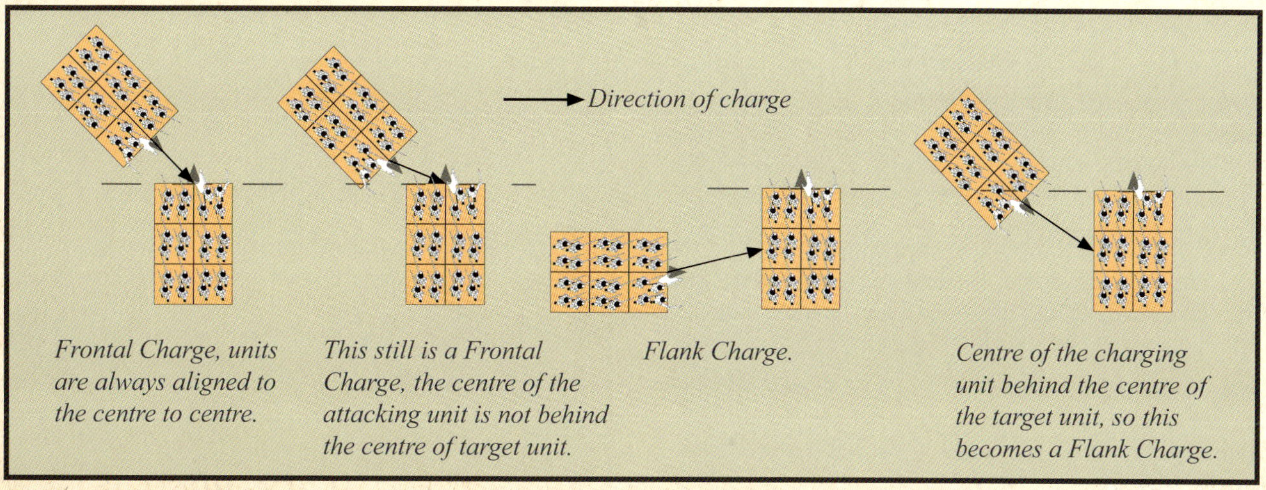

Deciding what is a Frontal or Flank Charge.

Moving Chargers

A successful charging unit should always be moved to the centre of the opposition front rank. This makes combat resolution more straightforward. If multiple units are charging the same enemy unit then they should charge to the nearest side of the enemy front rank.

To make a charge into a flank of a unit, the centre of the charging unit must fall behind centre front of the unit being charged. Likewise rear charges must be completely behind the rear of the unit being charged.

Charging units cannot change formation and only Light Cavalry can perform a 45° wheel during the first 8" of their move (this rule supersedes any national characteristic rules). Other units must move forwards in a straight line.

RUN!

The unit flees its perilous position by running with all possible speed towards its own table edge. The unit turns 180° and moves three times its normal move distance away from the enemy. The unit will finish the move facing towards its own table edge (with its rear facing the enemy).

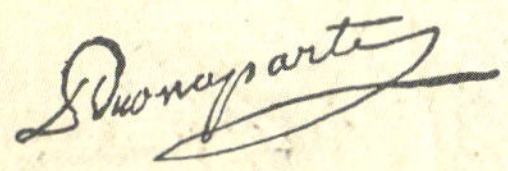

Running Distances:

Unit	Standard Movement Rate	Running Distance
Light cavalry	8"	24"
Heavy cavalry	6"	18"
Infantry	4"	12"
Artillery	4"	12"
Horse Artillery	8"	24"
Skirmishers	4"	12"

Running troops must roll 3D6 for fatigue; each 1 rolled on the dice causes a single fatigue casualty. If a running unit enters into or passes over any terrain then you must also roll separately to establish if there are any further fatigue casualties caused by the terrain feature. Fatigue casualties are not meant to represent troops actually dying of exhaustion, but rather troops who are no longer in a position to put up any resistance towards the enemy. Fatigue casualties are discussed in more detail on page 51.

If an artillery unit runs then the crews simply leave the guns and run for their lives.

Multiple charging units against a single unit.

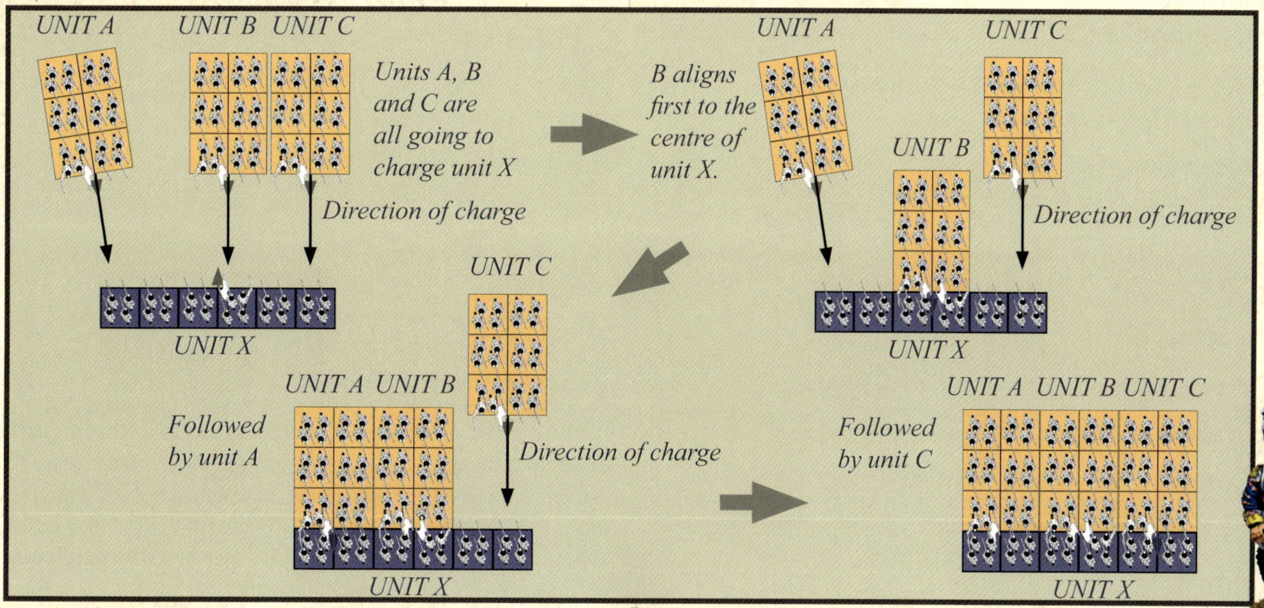

French Drummer.

Pursuit

Sometimes a unit will be forced to run as the result of a failed morale test. When this happens, the enemy unit that caused the runners to break and flee should give chase as fast as possible, moving its full charge distance in pursuit.

RETREAT AND FIRE!

The unit retreats half a standard Movement Rate away from the enemy while still facing them, then halts and fires off a volley.

Retreat and Fire Distances:

Unit	Standard Movement Rate	Retreat & Fire Distance
Light cavalry	8”	4”
Heavy cavalry	6”	3”
Infantry	4”	2”
Horse Artillery	8”	4”

Fixed artillery may not be given this command.

STAND AND FIRE!

The unit remains stationary and fires a volley at the enemy.

If the unit is being charged the firing is assumed to take place at point blank range.

If the firing unit inflicts two or more casualties the charging unit must pass a morale test to press the charge home. If the charging unit fails its morale test it halts 1” away from the unit on Stand and Fire!

With a Stand and Fire command it is assumed that artillery will always fire canister.

If a stand and fire order has been given as a result of Random Orders and there are no enemy units within fire range, arc of fire, or close combat then the unit will not waste ammunition and so will not fire.

RETREAT!

The unit retreats backwards, still facing the enemy, up to the following distances:

Retreat Distances:

Unit	Standard Movement Rate	Retreat Distance
Light cavalry	8”	8”
Heavy cavalry	6”	6”
Infantry	4”	4”
Artillery	4”	4”
Horse Artillery	8”	2”
Skirmishers	4”	8”

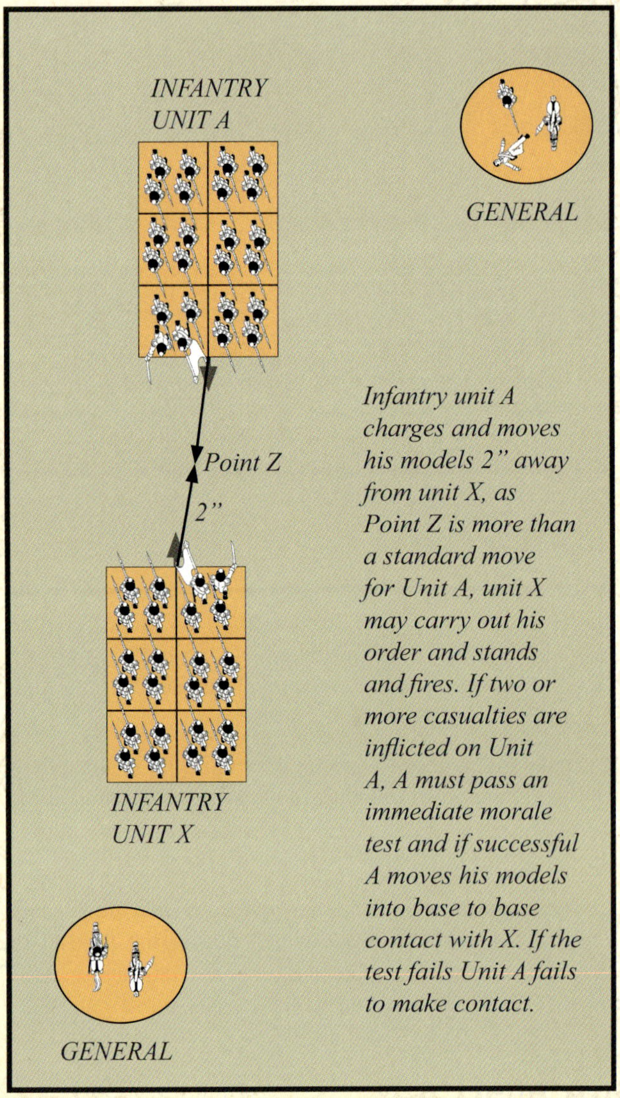

Stand and Fire.

With the exception of skirmishers retreating units may not fire. Skirmishers may fire immediately after completing their retreat move.

ADVANCE AND FIRE!

The unit moves its normal move distance towards the enemy, then halts and looses a volley of fire.

Advance and Fire Distances:

Unit	Standard Movement Rate	Advance & Fire Distance
Light cavalry	8”	8”
Heavy cavalry	6”	6”
Infantry	4”	4”

FORM LINE! FORM SQUARE! FORM COLUMN!

A unit can be ordered to change formation to line, square or column in order to better defend itself or become a more effective force on the battlefield. The unit given a formation order must remain stationary to perform the change of formation and cannot fire. Simply rearrange the unit bases on the spot to form the new formation.

Units Without Orders

All your units within engagement range **must** be given orders. If for some reason you forget, the following rules apply.

Infantry within engagement range of enemy cavalry automatically redeploy into square.

All other units within engagement range without orders should draw one Random Order card. Random Orders are only resolved after all other issued orders have been completed.

It really is a good idea to make sure you remember to issue orders to all your units in engagement range!

> After a little practice this part of the turn becomes very fast and great fun, as both sides have to decide what their most effective action or reaction should be.
>
> This also evens out some of the advantages that the player with the higher Initiative has. The player who goes first in the Tactical Orders phase will also have to move his men first in the Movement phase as his troops follow their orders, the player going second can therefore get some idea of what his adversary is planning.

French Light Infantry (Knötel).

French Tirailleurs (Knötel).

Unattached Units

If a unit becomes unattached it may not advance towards the enemy unless it gets an order to do so or a 'lost command' has occurred (see the Morale phase for details of lost command). If enemy troops enter the unattached unit's engagement range it will form a square (if the enemy unit is cavalry) or will retreat backwards (if the enemy unit is infantry). Once deployed into square formation an unattached unit will do nothing until the enemy cavalry move out of its engagement range.

Unattached cavalry units cannot form square and will always retreat if enemy troops enter its engagement range.

French Artillery Limber.

5. Movement

The whole art of war consists in getting at what is on the other side of the hill.

— Duke of Wellington

The player who won the initiative for the turn can choose to move one of the divisions of his own troops first or make his opponent move one of their divisions first – he might make his opponent go first to get an idea of what the enemy is planning. The player going second will then move all the units in his selected division. Play will then alternate and the process continues until both players have moved all their divisions and brigades.

Infantry, cavalry and limbered artillery can all move up to their full standard movement allowance unless any restrictions apply for terrain or weather. No unit can enter into the engagement range of enemy troops during the Movement phase. Once a unit moves within 8” of any enemy infantry (or 16” for light cavalry, 12” for heavy cavalry) it halts. The following turn the unit will need to be issued a command card order during the Place Command Cards Phase.

The movement rates for different troop types are:

Unit	Standard	March	Forced March
Light Cavalry	8”	16”	24”
Heavy Cavalry	6”	12"	18"
Infantry	4”	8”	12”
Artillery	4”	8”	12”
Horse Artillery	8”	16”	24”
Skirmishers	8”	16”	24”

Commanders can always move up to 24” in each Movement phase. They are not affected by any terrain or weather restrictions and cannot become fatigue casualties.

Deployed Artillery and Movement

Artillery pieces were usually moved around the battlefield by hooking the guns up to limber teams but when minor repositioning was needed crews could drag the guns around the battlefield by hand. There are two movement options open to unlimbered artillery:

1. Artillery crews can haul their gun forward or back up to 2” and can fire next turn.
2. Artillery crews can haul their gun forward or back up to 4” but may not fire next turn.

Simple Rules for Terrain

Terrain – the landscape of the battlefield, the trees, hills and rivers – of course has a huge impact on the movement of troops, as it does on other aspects of the battle. However, since the rules for terrain are necessarily complex, and since those rules affect many different aspects of the battle, we’ve presented the rules for terrain in a separate section, beginning on page 62. You might want to refer to that section to give yourself some idea of the effects of terrain on troop movements.

Alternatively, and particularly if you’re new to the rules, you might like to use the following simplified rules for terrain instead of the full version.

Terrain is divided into three categories: easy-going, difficult and impassable. You should decide which category each terrain piece belongs to based on its physical characteristics – how steep a hill is, how dense a wood is, how wide, deep or fast a river is and so on.

Easy-going terrain has no effect on units in column; units in other formations should roll a 1D6 and will suffer a single fatigue casualty on a roll of 1.

Difficult terrain can only be entered or crossed by units in column of march or attack. Units in column should roll 1D6 and will suffer a single casualty on a roll of 1. Units in other formations which are forced into difficult terrain for any reason suffer 1D3 fatigue casualties.

Impassable terrain cannot be entered or crossed by any unit. Any unit forced into impassable terrain automatically suffers a lost command.

These measures are, at best, a stop gap to get you going – try introducing the full rules of terrain once you’ve mastered the basics of orders, the turn sequence, combat and shooting.

Marches

Any infantry unit deployed in a column of march or column of attack formation at the beginning of the Movement phase, cavalry units in any formation, and any limbered artillery may perform a **march** move instead of their standard move. **Marching units may move double their normal movement distance**. In most cases a unit that has made a march move cannot then fire (see Fire Resolution on page 54). The notable exception to this rule is skirmishers (see the rules for Skirmishers on page 53).

Forced Marches

Infantry and cavalry that need to cover ground in double-quick time can perform a forced march. The unit must begin and end its move in column of march formation and it cannot wheel or make any formation changes. In this formation the unit is very vulnerable if caught by the enemy, but **it can move up to three times its standard move distance**. A unit on forced march cannot cross any terrain obstacles and must come to an immediate halt if it enters the engagement range of any enemy troops.

Roads

Column of March units performing a forced march along a road follow the same rules for forced marches, although the marching troops will automatically follow the curves and turns of the road without incurring any movement restrictions. If the units march into a junction they may follow whichever road they want to.

Limbered artillery can make a forced march along a road. It must start and end its move limbered up. Limbered artillery may only make a forced march on a road.

Troops making a march or forced march on roads never suffer any fatigue casualties.

Fatigue

The strain of a forced march, can cause troops to quickly become exhausted. Before moving the unit roll 3D6.

For each dice which scores 1, the unit suffers one fatigue casualty. Any unit suffering a casualty to fatigue (as with taking any casualties) must test its morale in the Morale phase.

Remember that fatigue casualties do not represent actual fatalities but rather troops who are unable to effectively fight on due to exhaustion. Fatigue casualties can have a terrible affect on the effectiveness of troops in battle and they should be carefully monitored.

Forced March Troops and Command

If a unit on a forced march comes within the engagement range of the enemy it automatically loses command and is subject to lost command orders (see Morale phase). The unit must make a morale test roll in the next Place Command Cards phase and all normal rules apply for the resultant orders.

Changing Formation

Throughout a battle units will need to change their formation for faster movement around the field or to deploy in a more effective attack or defence posture. Veteran, well-drilled troops can perform changes of formation faster than raw or Militia troops, who could become confused and disorientated by even simple formation changes.

Infantry

Infantry units (except for Guard and elite units, see below) must change formation at the beginning of the Movement phase before actually making a move.

Line infantry units can change formation and then make a standard move. The only exceptions to this rule are when the unit is forming or un-forming a square or forming a column of march. These formations take longer to achieve, so no further movement is possible that turn.

Militia units cannot move after making any formation changes; they spend the rest of the Movement phase bumbling around and trying to organise themselves.

All Guards and elite infantry units can change to or from any formation and then make a standard or march move, or make a standard move and then change formation.

Cavalry

Light and Guard cavalry units can change formation and then make either a standard or march move, or make a standard or march move and then change formation.

Heavy cavalry units can change formation and then make a march move, change formation and then make a standard move, or make a standard move and then change formation.

Militia cavalry units can only change formation and then make a standard move.

Artillery

Artillery can limber up and make a standard move or unlimber after a standard move. They can also haul their gun forward (or back) by hand up to 2" and fire next turn, or they can haul their gun forward (or back) up to 4" but may not fire next turn.

Horse Artillery

Horse artillery can limber up and make a standard or march move or unlimber after a standard or march move.

The British Army advances in columns of march.

POLITICAL SUBTRACTION.

From 5 Directors
Take 2
—
There remain 3 Consuls
From them take 2
—
And there remains 1 BUONAPARTE.

French Imperial Guard Grenadier (Knötel).

WHEELING

During the battle troops will not always want to move directly in a straight line, they will need to change their direction occasionally to avoid terrain features or to better position themselves to face the enemy. Changing direction in this manner is known as **wheeling**.

Moving units that are wheeling

Decide which direction a unit will wheel in. Units may not make a wheel in excess of the following angle:

Infantry can wheel up to 90º while making a standard move. Infantry can wheel up to 45º while making a march move. Infantry cannot wheel during a forced march (see page 51). If an infantry unit is making a forced march on a road it must follow the direction of the road.

Heavy cavalry can wheel up to 90º while making a standard move or up to 45º if on a 'march' move.

Light cavalry can wheel up to 90º on a standard or a 'march' move .

Artillery can wheel up to 45º and fire next turn, or up to 90º but may not fire next turn.

Turn the unit in the direction you wish to wheel. The unit should pivot on one of its front corners while the other corner will move around in the direction of wheel. The unit's outside edge may move up to the full move distance (either a standard move or a march move). The unit will remain in the same formation, so all bases will turn slightly with those on the outside edge turning the most.

Wheeling.

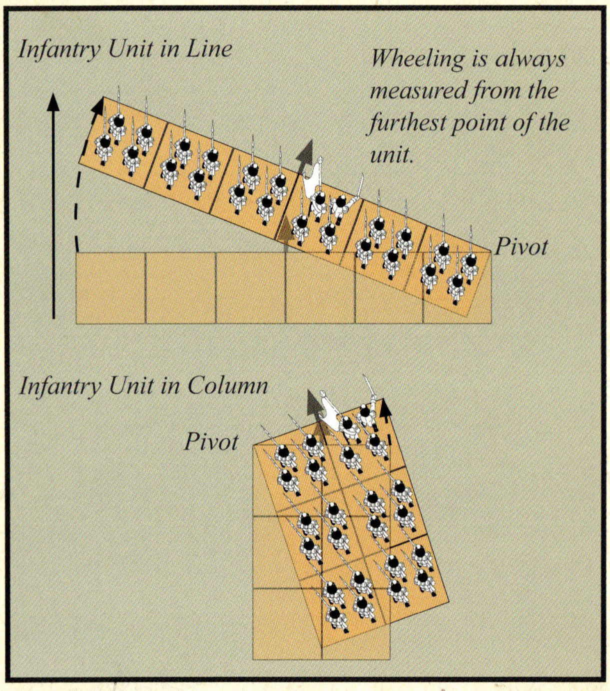

Normal rules for terrain apply while wheeling. A unit must not enter engagement range while wheeling; the unit will halt at the edge of engagement range instead ,wheel no further and halt.

LEAVING AND RETURNING TO THE BATTLEFIELD

The battlefield ends at the edges of the table; any units that leave the table while in the engagement range of enemy troops, or on lost command orders, are removed and take no further part in the battle.

In all other situations the leaving unit can make one attempt to return to the field. It must take a command test (see Test to Give Out Orders on page 45) and if successful the unit can return to the battlefield at the same point it left. If the test is failed the unit is removed and plays no further part in the battle.

6. Skirmisher Fire Resolution

Ability is nothing without opportunity.

— Napoleon Bonaparte

Light Infantry Companies

Many line units of the Napoleonic period included at least one light company – a small body of men which would detach themselves in a skirmishing group in front of their parent unit. We don't generally use miniatures to represent these skirmishers as doing so can make the battlefield look very cluttered and confusing (though if you wish to represent them, you'll find rules allowing you to do so on page 80). Instead, if a unit contains light companies it is assumed they provide the parent unit with extra firepower.

Light companies can make a standard or a march move and then fire, provided their parent unit is in line or column of attack formation. What this in effect means is that a line unit containing light companies (skirmishes) can move and still fire, albeit perhaps with less firepower than would normally be the case. The number of dice the light company may roll is detailed in the Army Lists. All light companies are armed with muskets unless otherwise specified in the Army Lists. Light companies and skirmishers have an increased range of fire, reflecting the fact that the unit's main volley fire was normally held back for firing at very close range.

Light companies are not restricted by the arc of fire rule and can therefore fire 360º all around. Measure the firing distances for light troops from the centre of the front, back or either flank of the parent unit thereby representing the fact that light troops can manoeuvre around their parent unit.

To decide whether the light company's marksmanship is effective, check their distance to the target, roll the number of D6 dice for light troops given in the appropriate Army List and check the chart below:

Musket Armed Light Companies

Firing Range	Distance	D6 Score Required to Hit
Effective	0-10"	5+
Long	10-20"	6

Rifle Armed Light Companies

Both the accuracy and range of the rifle was superior to that of the musket. If the rifle company's parent unit has remained stationary this turn, rifle armed Light Companies can re-roll any misses once.

Firing Range	Distance	D6 Score Required to Hit
Effective	0-12"	5+
Long	12-24"	6

Rifle Detachments

Some armies had a small number of rifle armed companies deployed within each division. Unlike other light troops, these operate completely independently of their parent unit and so are represented on the tabletop by models, as is the case for all other units. Each Army List details how many rifle detachments each division can contain, and how many bases of miniatures these consist of.

Rifle detachments must remain within 8" of at least one unit from their parent division at all times. The front of the forward most rifle base counts for the unit in terms of engagement range. If an enemy unit is over the Riflemen's standard movement rate away then, unless you give the unit another command card, the Riflemen will fire and then fall back to the nearest unit from their division and take up a position beside them. If the enemy unit is less than a standard move away and can be attacked the Riflemen move first, they will return straight to their parent division and take up a position alongside them. If the enemy moves first then the Riflemen are not moved.

Riflemen follow the same firing rules as light companies, except that distances are always measured from the centre of the rifle detachment itself rather than the parent unit. Rifle detachments can fire 360º all around.

Shooting at Rifle Detachments

Artillery batteries firing ball and other units firing volley at Riflemen should measure to the centre of the rifle unit. The Riflemen can ignore the first two casualties inflicted every phase. When enemy skirmishers or artillery firing canister target a rifle detachment, the firing distance is again measured to the centre of the Rifle Detachment, which can ignore the first casualty inflicted every phase.

Portuguese Cacadores (Light Infantry)

7. Volley Firing

If they want peace, nations should avoid the pin-pricks that precede cannon shots
— Napoleon Bonaparte

Units can fire a volley every turn provided they have made no more than a standard move.

Units making their first volley of the battle can march and fire, but cannot re-roll any misses. This supposes the units have entered the fray with their muskets already loaded and so are afforded the time to make a speedy march move before firing.

Arc of Fire

A unit can fire inside an area measured by drawing a line, 45º degrees to either side of its front bases. If the centre of the target lies outside this arc then the unit cannot fire at it. A firing unit should normally direct all of its firepower at a single target. If the arc of fire restriction makes this impossible the unit can divide fire between enemy units. If two enemy units are charging the firing unit, then fire can be divided between the charging units.

Nearest Target

Units must fire at the nearest enemy directly in front of them as these present the greatest immediate threat and are within the allowed arc of fire. If more than one enemy unit is the same distance away the firing unit must divide the dice roll score between the units (round up if fractions). Firing distances are always measured from the centre of the firing unit to the centre of the target unit.

Firing Formation

The amount of damage a unit can inflict upon the enemy is influenced by its formation:

Firing Unit's Formation	Number of D6 Unit Can Roll
Line	6 dice
Column	2 dice
Square	2 dice per side of square (the unit can only fire when on Stand and Fire! orders and only sides being confronted by the enemy can fire)

Guard units receive one extra firing die per unit.

Militia units must deduct one firing die per unit.

Cavalry armed with carbines roll a 1D6 to hit when in column of attack and 2D6 when in line formation.

Roll the appropriate number of dice and check the scores on the chart below. This indicates the **minimum score** needed to register a hit on the target unit. The greater the distance to an enemy unit, the harder they are to hit. Each successful hit causes one casualty.

Range and Score Required to Hit

Weapon	0-5"	5-10"	10-15"	15-20"
Musket	3+	5+	6	-
Carbine	5+	6	-	-

Opening Volley

Infantry firing their first volley of the game can re-roll any missed shots once, providing they did not march this turn and are not in square formation. It is a good idea to place a small tuft of cotton wool or some other marker next to units that have fired their first volley to help you keep track!

Stationary Units

Units remaining stationary – and therefore able to reload quickly – may re-roll any misses once in each round of firing. Running out of ammunition is a factor of Stationary Units volley firing.

Running Out of Ammunition

If a unit scores more than three 1s when rolling to hit it is indicating that it is running low on ammunition and can then only fire when on Stand and Fire! orders; to stock up on ammunition the unit must remain completely stationary for a whole turn.

Placing a model of an ammunition cart behind such units can help to remind you that they are replenishing their supplies. It also looks good!

Arc of fire, 45° degrees to either side front bases.

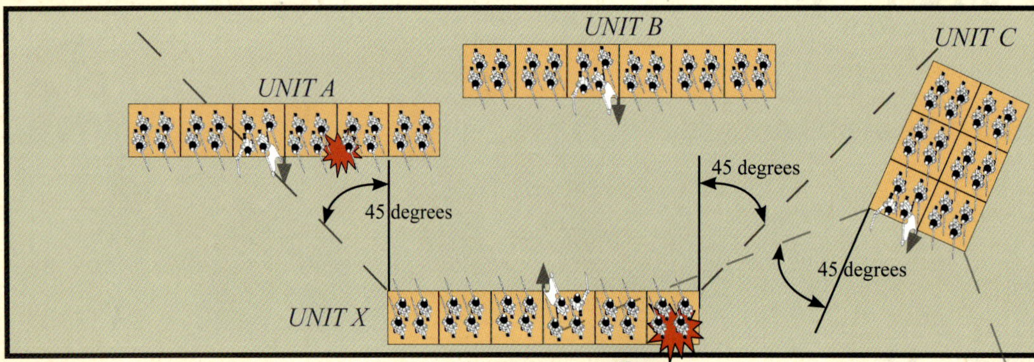

Unit A is targeted as it is the nearest target to X, For X, C is out of the arc of fire but for C, X is within the arc of fire and so can be fired upon by C.

8. Close Combat Resolution

A soldier will fight long and hard for a bit of coloured ribbon.

— Napoleon Bonaparte

Close combat occurs when two units become in base to base contact and attack each other at very close quarters using all the weapons at their disposal: swords, bayonets, musket butts and any other improvised weapons they can lay their hands on.

When enemy units on the tabletop are in direct base to base contact they count as being in close combat.

Units in close combat roll dice to determine who gains the upper hand in the fighting. The number of dice each unit should roll is determined by its formation and whether the unit is facing the enemy to its front, flanks or rear. This is shown in the chart below:

	Facing Target Unit's:		
Unit's Formation	**Front**	**Flank**	**Rear**
Artillery Battery	1	1	1
Light Cavalry Line	3	1	1
Heavy Cavalry Line	3(4)	1	-
Cavalry Column	2/2	1	1
Infantry Column	2/2	1	1
Infantry Square	2	2	2
Infantry Line	3(6)	1	-
Skirmishers	1	1	1

Units in column of attack formation who have charged more than their standard move into close combat can roll four dice and choose the best two scores.

Units in line formation roll 3D6 if confronted to the front.

If the line is in close combat with two or more units to the front then it can roll 6D6 dice for an infantry unit or 4D6 dice for a heavy cavalry unit. These dice must be split between the two opposing units.

Roll the appropriate number of dice and check the score of each die rolled on the chart below. This indicates the **minimum** score needed to register a hit on the enemy unit. The required score varies depending on the type of troops in the unit. Each successful hit causes one casualty.

Unit	Score Required to Hit
Infantry	4+
Heavy Cavalry	3+
Light Cavalry	4+
Skirmishers	6+
Artillery Crews	6+
Guards & Militia	
Guard Infantry	3+
Guard Heavy Cavalry	2+
Guard Light Cavalry	3+
Militia Infantry	5+

Exceptions

Some special rules apply in certain situations:

1. Troops who have force marched or who decided to Run prior to close combat require a 6 to score a hit with each die.
2. Troops attacking across a defended obstacle require a 6 to score a hit with each die.
3. Cavalry fighting an infantry square formation require a 6 to score a hit with each die.
4. Infantry not in square when facing cavalry require a 6 to score a hit with each die.

In cases where one unit is fighting multiple enemy units, divide the maximum number of dice (based on its formation) against each enemy unit. You'll have already seen that units in line formation receive extra dice in this situation, making it a particularly advantageous formation.

Russians and Bavarians in close combat.

Russian Infantryman.

Determining the Winner of Close Combat

The unit inflicting the greatest number of casualties upon the enemy wins the combat. Place markers on all the units involved to indicate the number of casualties suffered. In the case of multiple combats (more than two units on any side) the total hits of all the combats are added together.

Draws are possible (if both sides inflict the same number of casualties on each other), but the following special considerations apply:

Due to their massed ranks of men, a charging infantry unit in column which has charged more than their standard move distance counts a draw as a win.

Columns that have charged downhill also count a draw as win.

Charging cavalry who have moved more than their standard move distance downhill can also count a draw as a win.

If none of the above occur, the combat is considered to be undecided both units our moved directly back one inch facing each other.

Retreats

In the aftermath of close combat the victorious unit forces the defeated unit to retreat and can inflict further casualties upon them as they do so. The losing unit moves back a number of inches equal to the number of casualties it has suffered **but** remains facing the enemy. The victorious unit remains stationary.

Retreating Casualties

Further casualties can be inflicted on the defeated unit as it retreats, the number being determined by the victor's formation and its momentum as it entered combat. In the case of a multiple combat, retreating casualties are inflicted on each of the units by the combat winner. The winning player may choose which of his units in contact with each enemy unit to use when determining the number of extra casualties.

Infantry

Type of Formation	Number of Extra Casualties Inflicted
Charging Column	1D6
Stationary Column	1D3
Charging Line	1D3
Stationary Line	None

Square formations and column of march do not cause any further casualties.

Cavalry

Cavalry who charged less than a standard move distance cause 1D3 extra casualties while cavalry who charged more than a standard move distance cause 1D6 extra casualties.

Stationary cavalry cause no extra casualties.

Retreating Cuirassiers equipped with back as well as breast plates ignore the first extra hit.

Cavalry Breakthrough

Victorious cavalry units must immediately take a morale test. If they pass the test the cavalry unit can come to a halt, otherwise it moves an extra 8" (in the case of light cavalry) or 6" (in the case of heavy cavalry) in a straight line and suffers 1D3 fatigue casualties. The extra move could result in the cavalry catching up with the defeated unit and hitting them a second time, or even crashing into a fresh unit. The new bout of close combat that ensues must be fought immediately using all the above rules. This only happens following the first combat in each turn and there will be no cavalry breakthrough after subsequent combats. A player may voluntarily choose to move his cavalry unit forwards regardless of the result of the morale test, provided he is willing to suffer the associated fatigue casualties. Fatigue casualties do not count towards the following combat.

If any terrain features require the cavalry unit to make a fatigue test then this is calculated as normal. If the cavalry unit wins a second round of combat against the original defeated unit then the losing unit is destroyed and removed from the battlefield. If the combat is fought against a new unit, or the charging unit draws or loses the combat, the outcome is resolved as normal.

Cavalry Breakthrough.

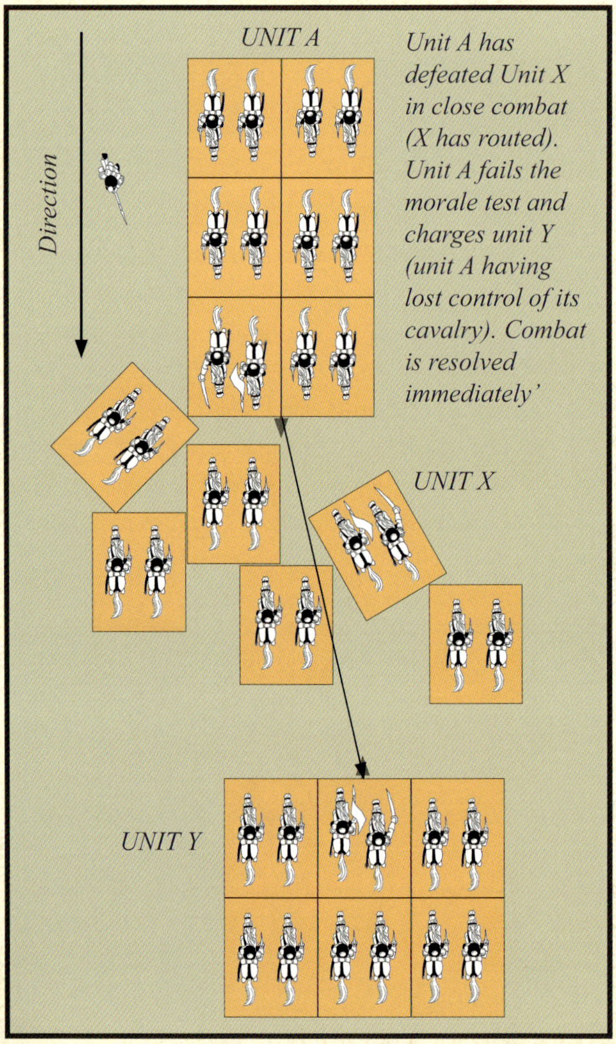

Square Formation

Squares can only fire if they have been given a Stand and Fire! order and only when the enemy unit is attempting to charge. Normal stand and fire rules apply with the exception that charging cavalry who have been shot at by a square are moved 1" away from the square. The cavalry unit must pass a morale test irrespective of whether any casualties have been inflicted. If the morale test is failed, the charge is halted (see failed charges, page 46). If the morale test is successful, the charge hits; move the cavalry into contact with the square.

Horses were very reluctant to charge into the thicket of bayonets presented by each face of the square formation. To reflect this cavalry always require a score of 6 when rolling to hit a square and no bonuses apply. Lancers can still re-roll misses if they charged more than their standard move distance.

Infantry in square formation fight as normal by rolling 2D6 and then applying any bonuses. If they are fighting cavalry and cause one or more casualties the infantry drive off the attacking horses, irrespective of how many casualties the cavalry may have inflicted. In effect, the unit is automatically victorious if they score one or more casualties, though the distance which they force enemy cavalry to retreat will generally be small.

If the square is unlucky enough to roll two 1s, or fails to score any hits at all and the opposing cavalry inflict two or more hits, the cavalry is deemed to have found a weakness in the formation and pours in.

A second round of combat is fought immediately, but this time the infantry unit is classed as being no longer in its defensive formation. Again the infantry roll 2D6, but this time they require scores of 6 to hit with no bonuses or special rules applying.

The cavalry roll to hit as normal, with any bonuses applying. If the infantry lose this bout of close combat they are destroyed and the unit should be removed from the battlefield.

Units in square formation may not pursue, consequently the opponent does not suffer additional casualties for retreating.

Cavalry Against Infantry

Infantry caught out in the open by cavalry and not in square formation roll the number of dice determined by their formation but they always require a 6 to hit, with any bonuses applied.

Cavalry fight as normal and if they win, the infantry unit is immediately destroyed and removed from the battlefield.

If the infantry are victorious they are deemed to have driven off the enemy cavalry who are pushed back as normal. The infantry then immediately deploy in square formation preparing themselves for any further engagements.

Artillery

If an artillery battery is beaten in close combat it is removed from the battlefield.

Skirmishers

If skirmishers are beaten in combat by any troops besides other skirmishers they are immediately destroyed and removed from the battlefield.

If skirmishers are beaten in combat by other skirmishers, the losing skirmishers are removed and the same number of casualties as they suffered are inflicted upon the parent unit.

Lancers

The charge of lance-armed troops could be devastating. To represent this, a unit of Lancers can re-roll any misses once as long as it has charged further than its standard move distance.

Prussian reserve or militia (Landwehr) lance armed cavalry.

9. Morale Tests

An army's effectiveness depends on its size, training, experience, and morale, and morale is worth more than any of the other factors combined.

— Napoleon Bonaparte

The morale of troops played an important role in the effectiveness of armies in the Napoleonic Wars. Some regiments, particularly poorly trained militia, would break and flee at the slightest provocation while others, notably veteran guards, would resolutely fight on to the death.

The cumulative affect of casualties sustained throughout the battle will have a direct impact on the morale of the regiment.

Morale tests in the morale phase always take place from the left edge of the battlefield of the player who has won initiative this turn.

Morale Tests

It is possible for a unit to take several morale tests in a turn.

All units that have suffered any casualties during the turn must take a morale test. When casualties are inflicted by artillery the test occurs at the end of the Artillery Fire phase.

The morale effects of a charge into close combat should be recorded as the event occurs. In the case of a charging unit a failed test stops the chargers from moving.

All morale test results are cumulative, so it is fairly common for units to rout after two successive morale failures.

Depending on their training, experience and the quality of their officers a unit should roll the following number of D6 to test for morale.

Troop Type	Number of D6 to Test Morale
Militia	2D6
Line Infantry	3D6 and discard lowest dice roll
Elite Infantry	3D6
Guard Infantry	4D6 and discard lowest dice roll
Militia Cavalry	2D6
Light Cavalry	3D6 and discard lowest dice roll
Elite Light Cavalry	3D6
Heavy Cavalry	3D6 and discard lowest dice roll
Elite Heavy Cavalry	3D6
Guard Cavalry	4D6 and discard lowest dice roll
Artillery	3D6
Guard Artillery	4D6

In order to pass its morale test the unit must roll a score more than the total number of casualties it has suffered during the battle. The number of casualties includes those caused by fatigue.

If the unit passes its morale test then it can fight on and function as normal.

If a unit fails its morale test it indicates that the Divisional General has lost command of the unit. The unit's officers desperately try to regain command and attempt to give orders. At this point the unit becomes subject to lost command orders.

There are only two forces in the world, the sword and the spirit. In the long run the sword will always be conquered by the spirit.

Austrian Army

Lost Command Orders

Lost command units roll 1D6 on the appropriate chart and react **immediately** as indicated:

Militia Infantry Lost Command Orders

D6 Score	Reaction
1	Run.
2-3	Retreat.
4-5	Remain stationary.
6	Unit receives a stand and fire order.

Infantry Lost Command Orders

D6 Score	Reaction
1	Retreat.
2-3	Remain stationary.
4-5	Stand and fire.
6	Advance and fire.

Elite and Guard Infantry Lost Command Orders

D6 Score	Reaction
1	Retreat.
2-3	Retreat and fire.
4-5	Stand and fire.
6	The unit charges.

Cavalry

D6 Score	Reaction
1	Run.
2-3	Retreat.
4-5	Remain stationary.
6	The unit charges.

Elite and Guard Cavalry

D6 Score	Reaction
1	Retreat.
2-3	Remain stationary.
4-6	The unit charges.

Artillery

D6 Score	Reaction
1	Abandon guns and run.
2-3	Limber guns and retreat.
4-5	Stay put but may only fire canister next turn.
6	Fire as normal next turn.

An infantry unit (including militia, line, guard and elite infantry) under lost command orders that finds itself confronted by enemy cavalry within engagement range must form square and remain there until it rallies. It can then redeploy with a command card if required or change formation freely in the normal way.

Multiple Morale Tests

If a unit fails a second consecutive morale test while already subject to lost command orders (a third consecutive failed test in the case of Guard units), then it automatically breaks and flees resulting in the unit to be considered being considered to be routing.

Rallying From Lost Command Orders

All lost command units have one attempt to rally on the following turn. The unit must take a morale test. If it passes, the officers of the unit have regained control; the following turn the units division general can give out a command card of his choice (, if required) or move or fire in the relevant game phases. If the unit fails its attempt to rally from lost command orders, the unit breaks and routs. Units quickly can switch between the lost command and rout!

Breaking and Routing

Breaking and Routing units move a full run distance **directly backwards**, which may mean that the unit suffers additional fatigue casualties dependent upon the terrain it encounters.

Any other friendly troops the fleeing unit comes into contact with should instantly receive a Random Command card. If the fleeing unit runs through a unit that is already on Lost Command orders then that unit also routs!

Rallying

All routing units have one attempt to rally on the following turn. The unit must take a morale test. If it passes the unit will stop fleeing and revert to lost command orders. If the test is unsuccessful the unit is deemed to have been permanently routed and should be removed from the table and will take no further part in the battle.

Russian Infantry.

10. Determining the Winner

Strategy is the art of making use of time and space. I am less concerned about the later than the former. Space we can recover, lost time never.

— Napoleon Bonaparte

Calculating Victory Points:

The last phase in each turn sequence is to calculate both armies' current Victory Points score; this is determined during this turn phase by each player adding together the percentage score value (PSV) of units which have either been destroyed (removed from the battlefield) or units that are currently routing.

Scores can increase or decrease every turn (and frequently do!) as units rout and reform, but as soon as it is found during this turn phase that a player has destroyed or routed 60% or more of the opponent's army (ie a player that has 60 or more Victory points) the game is over; it is then necessary to find out who has been victorious and, ultimately, the 'extent of the victory'. See Deciding the Overall Winner on page 35.

Some Players may wish to skip this breakpoint now and again and go to a new turn sequence either because the game being played is part of a particular scenario and the battle has to continue, or simply because it's obvious that no one has managed to destroy or rout more than 60% of the opponents' army. Some players may also skip this section just for fun to allow a General the chance to try and turn his defeat into victory – particularly if the battle has been one sided!).

Example:

Player A has a PSV score of 10 per unit

Player B has a PSV score of 11 per unit

Previous Turn

An army belonging to Player A has destroyed four units belonging to Player B (the units would have been at that point removed from the battlefield);

4 units destroyed multiplied by the PSV of 11 = 44 Victory Points

Not equal to or higher than the 60 required and so the battle continues.

An army belonging to Player B has destroyed five units belonging to Player A (the units would have been at that point removed from the battlefield);

5 units destroyed multiplied by the PSV of 10 = 50 Victory Points

Not equal to or higher than the 60 required and so the battle continues.

Berg Lancer of the Imperial Guard.

Current Turn

Player A has now also caused an additional two units from the army of Player B to rout

Player B has not destroyed or routed any further units of Player A.

At the end of this turn the score for Player A is now calculated as follows

4 units destroyed	4 X 11 = 44
2 units routing	2 X 11 = 22
Victory Points	**66** (66% of Player B's army has been destroyed or routed)
Player B still has	**50** Victory Points

Player A appears to have been victorious (Player A has destroyed more than 60% of Player B's army and has the most victory points) but this is not necessarily the case as points for the capturing of Colours, Eagles and Cannons need to be taken into account also. (See page 35).

Extent of the Victory

Example:

Player A:	
Victory Points from Player B units destroyed or routed:	66 (as above)
One Colour captured	5
One Artillery Battery destroyed in Close combat	5
Player A: Total Victory Points	76
Player B:	
Victory Points from Player A units destroyed or routed:	50
Four Colours captured	20
Two Artillery Batteries destroyed in Close combat	10
Player B: Total Victory Points	80
Difference in Scores (76 – 80)	4

Although the game stopped because Player A had more than 60 Victory Points he actually didn't win the battle as Player B had captured more Colours and army batteries than Player A.

To determine the 'extent of the victory' we need to check the table on page 35.

For a points difference of 4 the battle has been declared a 'draw'; in the spirit of the battle both armies effectively withdraw and regroup and live to fight another day!

The most important qualification of a soldier is fortitude under fatigue and privation. Courage is only second; hardship, poverty and want are the best school for a soldier.

— Napoleon Bonaparte

French Hussar 3rd Regiment (Knötel).

Terrain

When you have resolved to fight a battle, collect your whole force. A single battalion sometimes decides the day.

— Napoleon Bonaparte

Terrain is an important feature of any battle. Occupying key areas of the battlefield will lend a considerable advantage to your soldiers, while becoming mired in difficult terrain will massively hinder them. How to deal with the battlefield's terrain should be a key part of your strategy and something to carefully consider both before the game and when giving orders.

The Effects of Terrain

When placing pieces of terrain on the table before the battle starts, both players should agree on the effect each item will have on the troops. Some terrain may fall into more than one category and in these cases it is recommended that you apply the most severe of the related effects to the piece of terrain. In other cases it may be practical to combine effects – a hill on top of a wood, for instance, or a river with a very steep bank that itself counts as a sheer slope as the poor old footsloggers try and make their way up the other side!

Terrain affects all elements of the battle, from moving to shooting to close combat. For that reason we've collected the rules for terrain together here, rather than having them spread throughout the book (we have, however, noted key rules in the relevant section). It's a good idea to familiarise yourself with these terrain rules if possible, but they are not unduly complicated so when learning the rules you may prefer to simply get on with your battle and only refer to this section when necessary. We've summarised the effects of the different types of terrain below, and you can find detailed rules over the next few pages.

Hills

Gently Sloping Hills

Roll 1D6 for units carrying out a march, charge or run move on a gently sloping hill:

Score	Effect
1	The unit suffers a fatigue casualty and loses 3" of movement.
2-5	The unit loses 1D3" of movement.
6	No effect.

Steep Hills

Roll 1D6 for a unit carrying out a march or charge move on a steep hill:

Score	Effect
1	The unit suffers 2 fatigue casualties and loses 3" of movement.
2-5	The unit loses 1D3" of movement and suffers 1 fatigue casualty.
6	No effect.

Woods and Forests

Sparse woods

Units who march or charge through sparse woods must roll 1D6:

Score	Effect
1	The unit suffers a fatigue casualty and loses 1D6" of movement.
2-4	The unit loses 1D3" of movement.
5-6	No effect.

Woods

Units who march or charge through woods must roll 1D6:

Score	Effect
1	The unit suffers 2 fatigue casualties and loses 1D6 of movement.
2-4	The unit loses 1D3 inches of movement and suffers a fatigue casualty.
5-6	No effect.

Broken Ground and Scrubland

Units that march or charge through this type of rough or rocky terrain should roll 1D6:

Score	Effect
1	The unit suffers a fatigue casualty.
2-5	The unit loses 1D3" of movement.
6	No effect.

Marshland and Snow

Units that decide to charge through this type of terrain roll 1D6:

Score	Effect
1	The unit suffers a fatigue casualty.
2-5	The unit loses 1D3" of movement.
6	No effect.

British Highlander.

Rivers, Streams and Lakes

An army ought to only have one line of operation. This should be preserved with care, and never abandoned but in the last extremity.

— Napoleon Bonaparte

For the purpose of these rules, rivers and lakes are always considered to be un-crossable except at a ford or bridge. Streams are fordable at any point.

Infantry in column wishing to cross a stream or ford must move into contact with it and then stop. The following turn they cross the stream and the models are placed upon the other side but can move no further that turn.

Units in line must change to column formation before crossing.

Troops crossing rivers cannot fire.

Cavalry units making a standard move can cross a stream or ford as normal.

Cavalry who are on march or charge moves can cross over streams by 'jumping' them, but the unit must roll 1D6 and on a score of 1 they suffer one fatigue casualty.

Artillery can **only** cross streams or rivers via bridges or fords. It takes a full move for them to cross a ford. The artillery must halt when they come into contact with a river, and cross it in their next turn, in the same way as for infantry.

Bridges

Bridges are placed where a stream or river intersects a road; all roads which intersect a river must offer either a bridge or a ford to allow crossing. Units in column of march formation can make a standard move, a march move or a forced march move over bridges, without the need to halt; they may simply cross the bridge as part of their normal move.

Units in line and column of attack must change formation to cross a bridge.

Close Combat Over Bridges and Streams

All the usual rules for close combat apply with the following exceptions:

1. Infantry do not need to form square if defending, but count as if in square formation.
2. Bridges are too narrow to assault with lines or columns of attack. To get into combat, the charging unit fights as normal, except they only roll 1D6. The combat results are calculated in the normal way and stationary units roll dice in the usual manner.
3. Combat in fords and streams is very difficult. Units charging roll as normal, but require scores of 6 to hit.

In Saxony, lower Germany the French army defends a farmhouse; such building complexes can form major defensive strong points on the battlefield.

LINEAR OBSTACLES (HEDGEROWS, FENCES AND WALLS)

All terrain features that run in a continuous line but are possible for troops to clamber over, such as walls, hedgerows and fences, are described as linear obstacles.

Infantry units are slowed down by having to cross over linear obstacles. Their movement stops upon touching the edge of the terrain, and it then takes a full move to cross the obstacle next turn. If troops are caught while crossing an obstacle they count as being unformed. In the event that the crossing unit requires a command card, the crossing unit must reform or retreat back using the relevant order. If the order is failed or the crossing unit has no time to react it will require rolls of 6 to hit in close combat. While in close combat the unit cannot fire and if beaten must automatically roll for lost command.

Cavalry can decide to 'jump' the obstacle (unless enemy troops are directly behind the obstacle, in which case the cavalry unit must halt at the obstacle). If the cavalry unit decides to jump the obstacle a 1D6 is rolled and on a score of 1 the cavalry unit suffers a fatigue casualty.

Infantry can form up behind a hedgerow, wall or fence and, provided their bases touch the obstacle, can fire over it or through it. If half or more of the infantry unit's frontage is behind the obstacle, it ignores the first casualty in each phase.

Napoleon Bonaparte.

Linear obstacles are assumed to block line of sight so troops cannot charge at enemies behind them unless the enemy unit is positioned right up against it and is therefore visible. Troops charging at obstacles are moved directly into contact with the said obstacle.

In the case of walls, the infantry unit ignores the first two casualties in each phase.

An infantry unit completely behind a linear obstacle need not form square if charged by cavalry.

British Highland Infantry unit firing over a wall.

HILLS

The transition from the defensive to the offensive is one of the most delicate operations in war.
— Napoleon Bonaparte

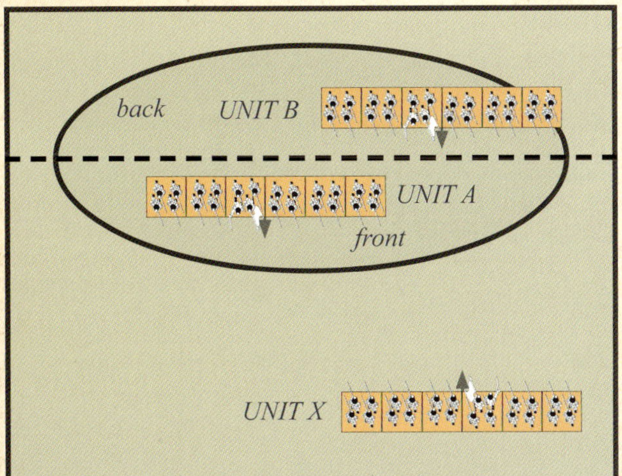

Infantry on a hill. Showing a line through the hill indicating front and back half. There are two units on the hill – unit A is on the front and so may fire at unit X, unit B is at the back, so may not fire at Unit X.

Hills vary dramatically in nature from gentle slopes and escarpments to sheer cliff faces. Players must agree on the severity of the hills (either steep or gently sloping) and delineate clearly on the battlefield where they start and finish.

All hills block line of sight and troops cannot fire or charge at enemy troops located on the other side. Units must be on the front half of a hill relative to the enemy in order to take part in firing or charging, but they can also be targeted by enemy fire and attacked in close combat. Units on the back half of a hill may not charge or fire and can only suffer casualties from howitzer fire.

The gradient, not the height, of hills determines the severity of the terrain. Other types of terrain, such as woods and buildings, can be on hills which gives a combined terrain effect.

Gently Sloping Hills

On hills with a gentle slope, units making a standard move can move as normal. If making a march, charge or run move then the unit's movement might become subject to restrictions. Charges can also be affected.

Roll 1D6 for units carrying out a march, charge or run move on a gently sloping hill:

Score	Effect
1	The unit suffers a fatigue casualty and loses 3" of movement.
2-5	The unit loses 1D3" of movement.
6	No effect.

Steep Hills

Steep hills reduce movement and cause fatigue in the units of troops climbing them.

Roll 1D6 for a unit carrying out a march or charge move on a steep hill:

Score	Effect
1	The unit suffers 2 fatigue casualties and loses 3" of movement.
2-5	The unit loses 1D3" of movement and suffers 1 fatigue casualty.
6	No effect.

Units cannot run up or down steep hills. They may attempt marches or charge (as well as normal moves) but must roll on the table above. Units may attempt a forced march, but must roll 2D6 on the table above and choose the lowest score.

Cliffs, Rock Faces and Mountains

Cliffs, rock faces and any other terrain possessing a sheer drop are impassable to all units and mountains can only be crossed at passes – all of which must be clearly defined and determined before the battle starts.

Austrian Jagers fight through a mountain pass.

Woods and Forests

In war there is but one favourable moment; the great art is to seize it!
— Napoleon Bonaparte

Wooded and forested areas are categorised as **sparse woods**, **woods** or **dense woods**.

All types of woodland block the line of sight, so you cannot fire through them at troops on the other side or at troops further than 2" in from the edge of the woodland.

Sparse woods

Sparse woods are generally made up of widely spaced trees with little or no undergrowth. Most coniferous woods are of this type.

Units who march or charge through sparse woods must roll 1D6:

Score	Effect
1	The unit suffers a fatigue casualty and loses 1D6" of movement.
2-4	The unit loses 1D3" of movement.
5-6	No effect.

Units within 2" of the edge of a sparse wood can fire and be fired upon in the normal manner. The unit in the wood ignores its first casualty from each phase.

Woods

The trees and undergrowth in woods are tightly packed reducing movement and potentially fatiguing the troops struggling to get through.

Units who march or charge through woods must roll 1D6:

Score	Effect
1	The unit suffers 2 fatigue casualties and loses 1D6 of movement.
2-4	The unit loses 1D3 inches of movement and suffers a fatigue casualty.
5-6	No effect.

Woods are impassable to artillery.

Units within 2" of the edge of the wood may fire and be fired upon in the normal manner. The infantry unit in the wood ignores the first two casualties in each phase.

Dense Woods

This type of woodland represents the once common areas of impenetrable, untouched forest that spread across Europe.

Movement in dense woods is impossible for all troop types except skirmisher detachments who treat the terrain as woods for movement purposes.

French light infantry skirmishers sneaking through dense woodland and Russian cavalry unit coming under fire.

Broken Ground and Scrubland

Units that march or charge through this type of rough or rocky terrain should roll 1D6:

Score	Effect
1	The unit suffers a fatigue casualty.
2-5	The unit loses 1D3" of movement.
6	No effect.

Artillery cannot move through broken ground or scrubland.

Units within 2" of the edge of the broken ground or scrubland may fire and be fired upon in the normal manner. The unit in the broken ground ignores the first casualty from each phase.

Marshland and Snow

The fate of a Nation may sometimes depend upon the position of a fortress.
— Napoleon Bonaparte

Marshland and areas covered in snow have identical effects upon units passing through them; neither type of terrain counts as cover. Artillery can pass through marsh and snow and units making a standard move are unaffected. If any type of unit makes a march move through this terrain, however, the march move counts as a forced march and all the usual forced march rules apply.

Units that decide to charge through this type of terrain roll 1D6:

Score	Effect
1	The unit suffers a fatigue casualty.
2-5	The unit loses 1D3" of movement.
6	No effect.

Buildings and Urban Areas

Buildings and built up areas on the battlefield can form a very important part of any commander's defensive position and tactics. At the Battle of Waterloo a major part of Wellington's strategy centred on holding the farm of La Haye Sainte and the walled complex of the Château of Hougoumont.

For ease of play, buildings in these rules are divided into three categories, each with its own rules: **single buildings**, **farms** (including hamlets and vineyards), and **villages**.

To enter any building or group of buildings being held by the enemy, attacking units must be on charge orders.

To avoid unnecessary complexity, we've limited the amount of interaction with buildings. In the rules presented here, you cannot fortify, destroy or set fire to buildings. However, if you wish to add this degree of detail to your games, you will find rules for such in the Advanced Rules.

Single Buildings

Single buildings can be a solitary structure or can consist of a few scattered smaller buildings covering an area approximately 6" by 6". Single buildings block line of sight for all firing and units cannot fire through them.

Only skirmishers can deploy inside single buildings, and even then only one unit of skirmishers may deploy inside at a time. It is assumed that the skirmishers inside will dig into the building making loopholes and firing positions so they can fire out as normal. There are no restrictions on arcs of fire.

Only skirmishers and infantry can assault single buildings.

Farms, Hamlets and Vineyards

Any collection of buildings representing a farm, hamlet, mill or any other small group of structures, often joined together by walls, fences and hedges, is referred to as a farm. Farms on the tabletop should have a maximum base size of 18" x 18".

A farm can contain up to one unit of infantry and a detachment of skirmishers, providing them with a very strong defensive position. It requires a full Movement phase to deploy into a farm; a unit may make standard move and then occupy a farm, provided their move is sufficient for the entire unit to reach it, but units making a march or forced march must halt at the farm's edge and may only occupy it on a following turn. A unit entering a farm cannot perform any other action on that turn other than to engage in close combat if necessary.

Troops deployed in a farm count as being under cover so the unit ignores the first two casualties it receives each turn. Half the infantry unit may fire in any direction, up to a maximum of six dice.

Units inside a farm are not subject to lost command orders; instead the unit is given a random command card but has the option of performing a Stand and Fire! order instead. After a second failed morale test (third in the case of Guards) the unit is removed from the battlefield.

Only skirmishers and infantry may assault farms. Cavalry and artillery cannot enter or occupy a farm, nor may they engage units within.

British deploy around a village

Prussian infantry to close for comfort to French Chasseurs, if they fail to redeploy into square formation they will face almost certain destruction.

Villages

A village is a large number of fairly densely packed buildings covering a minimum base area of 18" x 18". The players might need to delineate the edge of the village on the tabletop before the battle starts.

A village can contain two infantry units and two skirmisher detachments, or a single cavalry unit, or a single artillery battery. A full Movement phase is required to deploy into a farm; a unit may make standard move and then occupy a farm, provided their move is sufficient for the entire unit to reach it, but units making a march or forced march must halt at the farm's edge and may only occupy it on a following turn. A unit entering a farm cannot perform any other action on that turn other than to engage in close combat if necessary.

Cavalry and infantry may charge down any roads into or out of the village. Units cannot charge out of a village by any other means.

Troops in a village count as being under cover so the unit ignores the first two casualties it receives each turn. Half the infantry or artillery unit may fire in any direction, up to a maximum of six dice in the case of infantry.

Troops inside a village are not subject to lost command orders; instead the unit is given a random command card but has the option of performing a Stand and Fire! order instead. After a second failed morale test (third in the case of Guards) the unit is removed from the battlefield.

Only skirmishers and infantry can assault villages.

Towns

For the sake of simplicity towns should be divided into sections made up of single buildings, villages and farms. Villages can represent the more built up areas, with single buildings serving as strong points or towers. Always leave clear areas such as parks so that some unhindered combat can take place. It is very important that all players are clear as to exactly what each part of the town is represented before the battle begins, otherwise it will all become very confusing!

There are no siege rules in the game as this is an entirely different form of warfare to that fought by massed troops on a battlefield. If you do want to attack curtain-walled towns with your field armies, who possess no siege cannon, here are a few suggestions.

The town walls cannot be destroyed so attacking troops can only enter via gates. Use the rules for bridges to represent this but treat the defenders as standing behind linear obstacles.

Infantry can deploy in line along walls but roll a maximum of three dice in combat. All other rules for troops in farms apply to troops in fortifications. Towers can contain a single cannon or a detachment of skirmishers. Once again the players must decide before the battle starts details such as where gates are positioned and which buildings are classified as towers.

The French attack a Russian walled town!

The spectacle of a field of battle after the combat, is sufficient to inspire Princes with the love of peace, and the horror of war.

— Napoleon Bonaparte

French Line Lancer (Knötel).

Napoleonic Battles

French Cuirassiers fall upon British cavalry; Heavy Dragoons, Light Dragoons and Hussars.

Gallery

Prussians deployed in squares, prepare to receive French Chasseurs and Lancers.

Russian Cuirassier.

Bavarians, France's ally, until Napoleon's Empire started to crumble and they switched sides and joined in with Allied armies.

Prussians and Austrians march forward, but French horse artillery provides long range firepower. Will this Allied onslaught fail against such determined defenders?

French Dragoon Officer.

Three allied divisions, tightly deployed advance. While mutual support amongst units is often the key to victory, units deployed too close can become congested reducing the numerical advantage of this army.

This one will take careful timing, when defending it's easy to lose control of the battle allowing your adversary to gain the upper hand.

Bavarian Chevauxleger.

The Napoleonic Wars saw many massive cavalry versus cavalry encounters; here the French General has deployed his Cuirassiers and Guard Cavalry interposed with fast moving Horse Artillery! British Horse Artillery can be seen retiring on the left.

British Light Dragoon Officer.

GALLERY

The British Army marches through Spain in the Peninsular War. The Duke of Wellington himself questions a local about the lay of the land, the Spaniard seems to have had too much of the local Rioja *to care!*

British heavy cavalry surge forward towards their infantry, who under pressure from French Cuirassiers have deployed in squares.

Portuguese Cavalryman.

French cavalry versus Prussian infantry part 1. *French lancers and Chasseurs prepare to launch themselves upon Prussian infantry columns; the Chasseur unit in front of the artillery battery will suffer huge loses from canister fire at this close range.*

British Heavy Dragoon.

French cavalry versus Prussian infantry part 2. It was a trap! Prussian cavalry previously hidden in the woods threaten the flank of the French cavalry, who quickly have to switch their attack to this new danger.

This time the prudent Prussian general has formed his infantry into defensive squares, whilst very good formations against cavalry, squares are vulnerable to artillery fire.

Prussian Reserve Lancer Officer.

In war, as in politics, no evil - even if it is permissible under the rules - is excusable unless it is absolutely necessary. Everything beyond that is a crime.

— Napoleon Bonaparte

French General of Cuirassiers (Knötel).

ADVANCED RULES

The moral is to the physical as three to one.

— Napoleon Bonaparte

After you have fought a few battles using the basic rules, you might want to introduce some extra details into your games to add realism and increase the number of strategic options open to players. Feel free to use as many or as few of the Advanced Rules as you want, just remember to make sure that all players agree before the game starts which part or parts of the Advanced Rules you are using. It is recommended that players who want to fight very large battles within a limited amount of time use the basic rules only.

For games that form part of a campaign, or are based directly on an historical battle, you might need to add some special extra rules of your own to make the game fit the scenario. This is fine as long as all players are aware of and agree the changes to the rules before the game starts.

LIST AND SUMMARY OF ADVANCED RULES

Commander in Chief's Intervention

Generals as Casualties

Light Infantry & Skirmishers

- Representing Light Infantry
- Deploying Skirmishers
- Skirmisher Firing
- Skirmishers in Close Combat
- Riflemen and Jaegers

Deployment

- Reserve Troops
- Flank Attacks

Grand Battery Fire

Rockets

- Range, Nearest Unit & Line of Sight
- To Hit
- Misfires

Battalion Guns

Setting Fire to Buildings with Howitzer Shell Fire

- Single Buildings
- Farms
- Villages
- Putting Out Fires
- Evacuating Buildings
- Rockets
- Destroying Bridges

Engineers and Pontoon Wagons

- Before the Battle
- Blowing Up Bridges
- Pontoon Trains

Friendly Fire

Movement Rates for Squares

Back to Back Fighting

Surrendering

Weather

- Ground Condition
- Air Condition
- Ground Condition
- Atmospheric Condition

French Infantryman 1805 (Knötel).

GENERALS

Every soldier carries a marshal's baton in his pack

— Napoleon Bonaparte

COMMANDER IN CHIEF'S INTERVENTION

Once per turn the Commander in Chief can intervene in the actions of one of his subordinate Generals while they are giving out orders. If the Commander in Chief is within 24" of a unit that fails a command test, that unit can re-roll using the Commander in Chief's command rating. The unit must accept the result of this second attempt. If it passes then the order can be carried out after all. If the unit fails then the normal rules for failed command tests apply, but the result should not be counted towards future morale tests.

If the unit is between 24" and 36" from the Commander in Chief it can still try to re-roll, but his command rating is reduced by two, it being assumed that the commander has sent out an Aide de Camp to act on his behalf.

Once per turn a Commander in Chief can attempt to rally a unit on lost command orders. The unit must be less than 24" away from him and should test its morale again, this time adding an extra 1D6 to its morale test score. If the unit fails this special morale test the result should not be counted towards future morale tests.

GENERALS AS CASUALTIES

In the basic rules, generals cannot be captured or killed; with this advanced rule they can. If a general is within 2" of a unit that receives casualties for any reason other than fatigue the player should roll 1D6. A score of 1 indicates that the general was caught up in the fighting and has been slain.

Whenever a unit routs and is removed from the table, roll 1D6 to see if the general of that division is involved. A score of 1 shows that he has been killed. Similarly if a unit is defeated in close combat roll 1D6; a score of 1 results in the division's general becoming a casualty or is captured.

For any Divisional General that is lost, a subordinate officer is assumed to immediately take command. You should roll for this new commander's command rating in the usual manner, but deduct 2 from the final score. This replacement commander can himself be killed by later actions, using all the rules above.

Read over and over again the campaigns of Alexander, Hannibal, Caesar, Gustavus, Turenne, Eugene and Frederick... This is the only way to become a great general and master the secrets of the art of war.

— Napoleon Bonaparte

The Emperor Napoleon Bonaparte

Napoleon was born in Corsica in 1769 (Napoleon is far from the only empire-builder to be born at his homeland's fringes and his Corsican origins are reflected in his Italian-influenced name). Napoleon trained as a military officer and rose to prominence during the French Revolution before staging a coup d'état in 1799, taking the title First Consul. Military gains over the early years of his reign would result in the birth of the first French Empire and Napoleon's coronation as Emperor. Imperial France would dominate Europe for most of the next decade.

1812 and the decision to invade Russia proved a defining moment in Napoleon's life and reign – the invasion was a catastrophe, his army all but destroyed and his reputation shattered. Just a year later, the so-called Sixth Coalition of European powers invaded France, defeated Napoleon and exiled him to Elba. He would return, briefly, in 1814, supported by a hard core of his most loyal supporters before being finally defeated at Waterloo in 1815 and exiled to St. Helena where he died (allegedly poisoned by his captors) in 1821.

Arthur Wellesley, Duke of Wellington

Wellesley was not born a Duke, but instead born in Ireland in 1769 (the same year as Napoleon, 3 months Wellington's junior) to an untitled family and commissioned into the army as an ensign. He became a colonel in 1796, became a field marshal in 1813 and acquired his dukedom in 1814 (he was thus known as Wellington by the time of his most glorious victory at Waterloo in 1815, but not during the Peninsular War where he earned his fame).

Wellington (then Wellesley) was sent to Portugal in 1807 and experienced early successes against the invading French, though his career nearly came unstuck when he was (wrongly) accused of approving the Convention of Sintra which permitted, indeed even aided, the defeated French army in withdrawing from Lisbon. The move against Wellington was a political one and he was eventually cleared, returning to the peninsula in 1809. Wellington was incredibly popular with his own men, particularly the lower ranks, a fact which no doubt aided in keeping morale high in what proved to be a long war.

LIGHT INFANTRY & SKIRMISHERS

Light infantry are a specialist infantry troop type. They're mainly ignored in the basic rules (though are represented by the rules for light companies) because of variety and additional complexity they bring to games. Once you're familiar with the rules, you should have no problem at all incorporate light infantry into your battles.

Many armies of the Napoleonic Wars included some light troops amongst their units and you can find details of these in the Army Lists.

Light infantry units represent troops who have been trained with the ability to detach large numbers of their formation from the unit to act as skirmishers. Famous examples of these troops include the British 60th and 95th Rifles, the Portuguese Cacadores, French Legere units and all Jaeger units. Some of these regiments were partly or wholly armed with rifled weapons.

Light infantry units can fire and fight to the same level as standard infantry units; because of their high level of training most are actually higher quality soldiers than their

Light Infantry L to R, French, Austrian & Portuguese.

line counterparts! Light infantry can also be classed as Elite or Guard. The terms 'light infantry' and 'skirmishers' are both used – in practice they mean the same thing. We often use light infantry when referring to the troop type and skirmishers when referring to the actual bases of miniatures, but there's no real difference and the terms are completely interchangeable.

REPRESENTING LIGHT INFANTRY

Light infantry units require additional bases if they are deployed as skirmishers; players should only have two miniatures per 40mm x 40mm base (see page 20).

When you wish to deploy skirmishers, these light infantry bases will be placed on the tabletop, like other units. These bases do not represent a literal replacement of the unit's formed troops into skirmish order, they merely indicate that skirmishers have been deployed, so when you place skirmishers on the tabletop do not remove any bases of models from the parent unit.

DEPLOYING SKIRMISHERS

A unit can deploy a certain number of skirmishers as detailed in the Army Lists. If a unit wishes to deploy skirmishers, it can only make a standard move that turn. As the skirmishers detach they can move an additional standard move distance measured from the front of the unit. The skirmisher line can deploy in front of its parent unit or any other unit in its division.

When a unit deploys skirmishers, these will count as part of the unit for determining engagement range. An enemy unit is within engagement range if it is within engagement range of *any* of the unit's bases, be they skirmishers or formed-up bases in the parent unit. If an enemy unit is over the skirmishers' standard movement rate away then, unless you give the unit another command card, the skirmishers will fire and then fall back into their parent unit.

Skirmisher detachments deployed.

Skirmishers can be given a different command to their parent unit.

For example, if you gave the skirmishers Stand and Fire! orders, they will fire and retreat. If the skirmishers are given Advance and Fire! orders they will fight as per the normal rules for that order.

If the enemy unit is less than a standard move away and the light infantry move first, the skirmishers will return straight to their parent unit immediately, but if the enemy moves first then the skirmishers stay where they are.

Skirmisher Firing

Skirmishers can fire 360° all around and so are not restricted by the arc of fire rule.

Artillery batteries firing ball and other units firing volley at skirmishers should measure to the front (centre) of the parent unit. The light infantry unit can ignore the first two casualties inflicted every phase. When enemy skirmishers or artillery firing canister target a light infantry unit, the firing distance is measured to the centre of the skirmisher's base, who can ignore the first casualty inflicted every phase.

Casualties are always inflicted on the parent unit as it is assumed casualties in the skirmisher line will be immediately replaced.

The parent unit, or any other friendly unit, can volley fire through the skirmisher line as long as the skirmishers are no more than 2" away. The effect of shooting will be reduced, however; the number of skirmishers the parent unit has deployed will affect how many dice are rolled when firing a volley (see the chart below).

The normal arc of fire rule applies to the parent block.

Number of bases deployed as Skirmishers	Reduction to Firing Dice:	
	Column of Attack	Line
1	2	None
2	2	1
3	*	2
4	*	3

*If three or four bases are deployed as skirmishers, the parent unit can no longer deploy in column of attack.

Units behind the skirmisher line can ignore the first casualty caused by ball shot from cannon and from infantry firing volleys.

Skirmishers in Close Combat

Skirmishers always require a score of 6 to inflict casualties in close combat and only ever cause extra casualties after winning a round of close combat against other skirmishers or artillery crews. If the skirmishers draw or lose a round of close combat, their unit automatically loses **all** its skirmishers and any casualties inflicted on the skirmishers are transferred to the parent unit.

Riflemen and Jaegers

Many nations used rifle armed troops; the various German states recruited foresters who were already excellent shots before joining the army. In German speaking armies, including those of Prussia and Austria, these riflemen were given the title Jaegers (the German word for hunter), a title which was also used to describe the riflemen in Russian armies. Light infantry units who were primarily armed with muskets would often include a number of rifle armed companies.

Light infantry deployed in skirmish order are presumed to be equipped with rifles. The only exceptions are French and Bavarian skirmishers, who are always musket armed.

A good example of a well deployed and well balanced army.

DEPLOYMENT

Two armies are two bodies which meet and try to frighten each other.

— Napoleon Bonaparte

Generally, all divisions and units must be deployed on the tabletop before the battle starts. The rules presented here give you the option to leave divisions in reserve (off-field reinforcements who will arrive during the battle) or to try and use the classic Napoleonic tactic of flank attacks.

RESERVE TROOPS

You can deploy one or more divisions as a reserve. These units are still part of your army (so they will count towards your infantry PSV) but they begin the game off the battlefield and should not be placed on the table at the start of the game. Divisions in reserve cannot take any part in the action until they have been ordered to advance onto the battlefield; this occurs at the beginning of the Movement phase when the player must inform his opponent that he is trying to deploy his reserves.

The reserve division must pass a command test against the Commander in Chief's command rating and, if successful, will deploy onto the battlefield the following turn.

Reserve troops can deploy anywhere along the player's own table edge provided there are no enemy models within 16".

All reserve division units, with the exception of light cavalry, should be deployed in a column of march formation. The whole division, including artillery, is deployed in a single turn, irrespective of its size and cannot fire or change formation during its deployment turn.

Light cavalry cannot charge in the turn the division is deployed but can assume any formation they wish.

If you are fighting on a *very* large table then you should divide it into 6 foot sections before the battle starts and make a note of which section each reserve division will arrive in.

FLANK ATTACKS

Flank attacks are dealt with in much the same way as reserve troops except that the command test to deploy is taken against the Divisional General's command rating rather than the Commander in Chief. The Divisional General's command rating is reduced by two for this test.

Units involved in a flank attack don't count towards your PSV until they arrive on the battlefield. If they pass their command test you can choose to deploy your troops on your opponent's side of the table or your own. They should be placed no further than 16" from the table edge and no closer than 2" away from the enemy.

Units that have been flank attacked can attempt to react to the enemy's sudden arrival; the unit must take a command test and if successful can turn and face the enemy, even if it has already moved this turn.

Deployment Formations Examples.

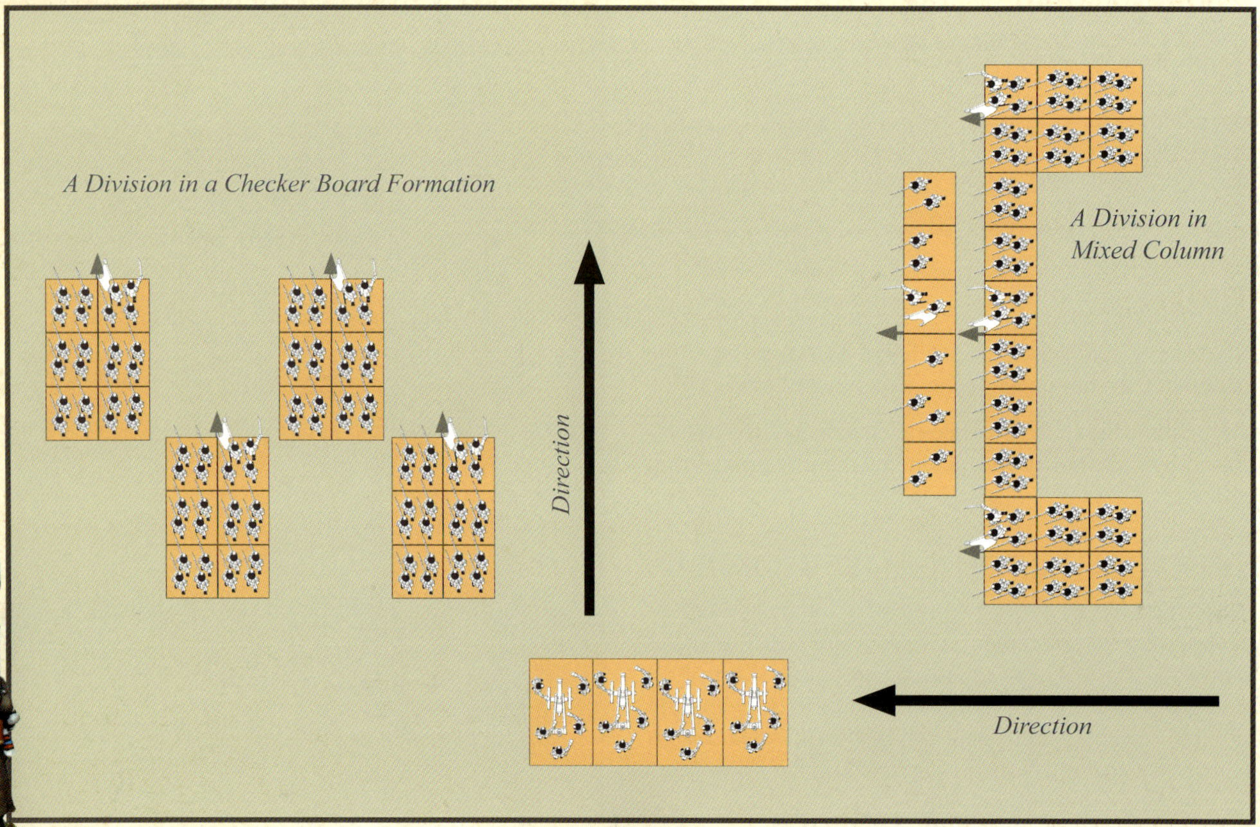

Russian Infantryman.

Grand Battery Fire

Concentrating artillery into massed batteries that could blast holes in the enemy was a favourite tactic of the age; massed battery fire could inflict severe losses and was capable of breaking whole divisions apart.

Two or more batteries are needed to form a grand battery and must be deployed so that there are no gaps between the bases of the component units. A grand battery may contain a maximum of four batteries.

When combining to form a grand battery, artillery units can detach themselves from their Divisional General's command and use the Commander in Chief's command rating.

Once formed, the grand battery counts as a single unit for firing purposes. Morale is still calculated for each unit separately but all tests are taken with an additional 1D6 added to the score. When firing ball shot – supported by any howitzers – the whole battery must fire at the same enemy division, but not necessarily at the same target unit, thereby making it possible for the grand battery to target multiple units.

Each additional battery will give the grand battery a plus 1 to hit, up to a maximum of plus 2 (thus a grand battery of two batteries adds +1, a grand battery of three, or even four, grand batteries adds +2).

The effect of canister shot is calculated for each unit individually as if two guns either side of the unit are ready to fire on Stand and Fire! orders.

Rockets

British and Austrian armies employed a number of rocket troops that fired rockets from small portable launch platforms or larger ones fixed to carriages like a cannon.

Mounted rocket troops fight as light cavalry but they cannot fire their rockets if they are within engagement range of enemy troops.

Rockets are artillery and unless stated otherwise follow all the normal rules for artillery.

Range, Nearest Unit & Line of Sight

Rockets can fire up to 24" during the Artillery Fire phase and the rocket unit can then move as normal afterwards. Rockets do not require a line of sight to their target, nor are they required to target the nearest unit.

Rockets fired from carriages have a range of 48". All other rules are the same (no need for line of sight, etc).

To Hit

Each rocket unit rolls 3D6 and all three dice must score a hit to be effective. Rockets fired from carriages hit on rolls of 5 or 6.

If successful roll 1D6 and then as many dice as the 1D6 score; any scores of 4, 5 or 6 will cause a casualty (1,2 or 3 has no effect).

If you are firing at a cavalry unit and score any hits at all (even if not the three numbers required to inflict casualties) the horses panic and the cavalry unit automatically goes on lost command orders, (which are rolled for at the end of the Artillery Fire phase). If the unit is already on lost command orders they will rout.

Misfires

Rockets can be unpredictable and sometimes as dangerous to the launchers as to the target. When firing, if three 1s are scored, the rockets have exploded prematurely during firing. Roll 1D6 then roll as many dice as the 1D6 score; any scores of 4, 5 or 6 cause a casualty on the rocket firing unit.

Despite their inherent unreliability rockets were very good at setting fire to buildings and causing panic among horses.

Battalion Guns

During the eighteenth century most infantry battalions and regiments contained a small detachment of very light guns. These gradually disappeared during the Napoleonic Wars (although Napoleon often used captured enemy guns) as the destructive power of massed artillery increased. One gun and four crew models can be used to represent battalion guns, and for firing they count as horse artillery. Movement is the same as for foot artillery and all normal rules apply.

Battalion gun emplacements have no separate morale as they are a detachment of their parent unit and their morale and reactions are accordingly based on their parent unit. If the parent unit charges or gets caught up in close combat, the gun crew do not get involved, leaving the fighting to the infantry. Battalion guns must remain within 2" of their parent unit at all times.

Setting Fire to Buildings with Howitzer Shell Fire

Howitzers can attempt to set fire to any buildings. The whole artillery battery must fire at the same building and any hits from cannon firing shot are resolved in the normal manner.

Players should always decide which buildings are wooden before the battle starts. For the purpose of these rules all buildings in battles of the Russian campaign are assumed to be wooden unless players agree otherwise.

Single Buildings

A single hit from a howitzer on a wooden building starts a small fire. Units inside a wooden building can either attempt to extinguish the fire or evacuate. On the turn that a building is first set on fire, these are the only options available to units within – on subsequent turns they may act as normal if they wish, though they may suffer the consequences.

If both firing dice give hits on a wooden building then it is instantly set ablaze. Any units inside must evacuate.

Non-wooden structures require a direct hit to set them alight. Troops inside can either attempt to put the fire out or can evacuate, but cannot do anything else in the turn during which the blaze begins.

On future turns, units can opt not to extinguish the flames and are free to operate in the normal manner, but an untended fire will smoulder away and any further direct hits on the building will set it ablaze. Once a building is ablaze, units inside must evacuate.

Farms

The same rules apply as for single buildings, but wooden structures require two hits to start a fire, non-wooden structures require three. Once alight, further direct hits will set the farm ablaze.

Villages

The same rules apply as for single buildings, but wooden structures require three hits to start a fire; non wooden structures require four hits. Once alight, further direct hits will set the village ablaze and units must evacuate.

Putting Out Fires

Units inside farms or villages can put out fires at the rate of one fire per unit as long as they do not do anything else in the same turn. Artillery or cavalry must leave any structures which are on fire. Until the farm or village is completely ablaze troops inside can continue to fight in the normal manner.

If you are using the weather rules and the battle is in heavy rain only wooden buildings can be destroyed by fire.

Evacuating Buildings

To evacuate a building, a unit must leave it in a direction away from any enemy units. If the building is completely surrounded the unit automatically surrenders and is removed from the battlefield. The building is destroyed.

Rockets

Any hits, including glancing ones, from rockets automatically set fire to any structure in the same way as a 1 hit from howitzers. Full rules for rockets can be found on page 83.

Destroying Bridges

Wooden bridges can be set alight with howitzer shell fire. Each hit causes a fire, and three fires burning simultaneously will completely destroy the bridge. Each unit on in touching the bridge can put out a single fire per turn but cannot do anything else on the same turn. Only infantry or skirmishers can cross a bridge that is on fire. Brick or stone bridges can only be destroyed by engineers (see page 86).

French Guard Horse Artillery howitzer.

Engineers and Pontoon Wagons

Unit of French Engineers

All armies employed engineers and pontoon trains of varying quality. In these rules armies can contain one unit of infantry engineers for every four infantry divisions.

Engineer units cost 25 points and consist of six bases of engineers. Engineers may fire a single musket shot in volley fire if not performing any engineering tasks. They will not charge into close combat and if charged and forced into close combat they count as skirmishers. Morale tests are made with 2D6 dice.

Before the Battle

Before the battle your engineers can destroy any one bridge that is within your army's deployment area or strengthen the defences of any one building, farm or village in your deployment zone; in a reinforced building, the defenders can ignore an extra casualty each phase.

Engineers can also construct field fortifications by digging earth ramparts and using gabions and wood; these were mainly used for siege warfare but occasionally appeared in set battles.

To build earthworks before a battle, the player must be fighting in a defensive role, a role which will only usually occur in a scenario driven battle or one based on an historical event; earthworks were used to great effect at the Battle of Borodino, for example, where a great redoubt and smaller arrowhead defences were constructed.

Earthworks count as linear obstacles. Each unit of engineers can build earthworks large enough to hold two artillery batteries, or one infantry unit in line or two infantry in column of attack (in essence, this means the earthworks will have a frontage roughly equivalent to six infantry bases).

Map of the Siege of Badajoz, Spain, 1812.

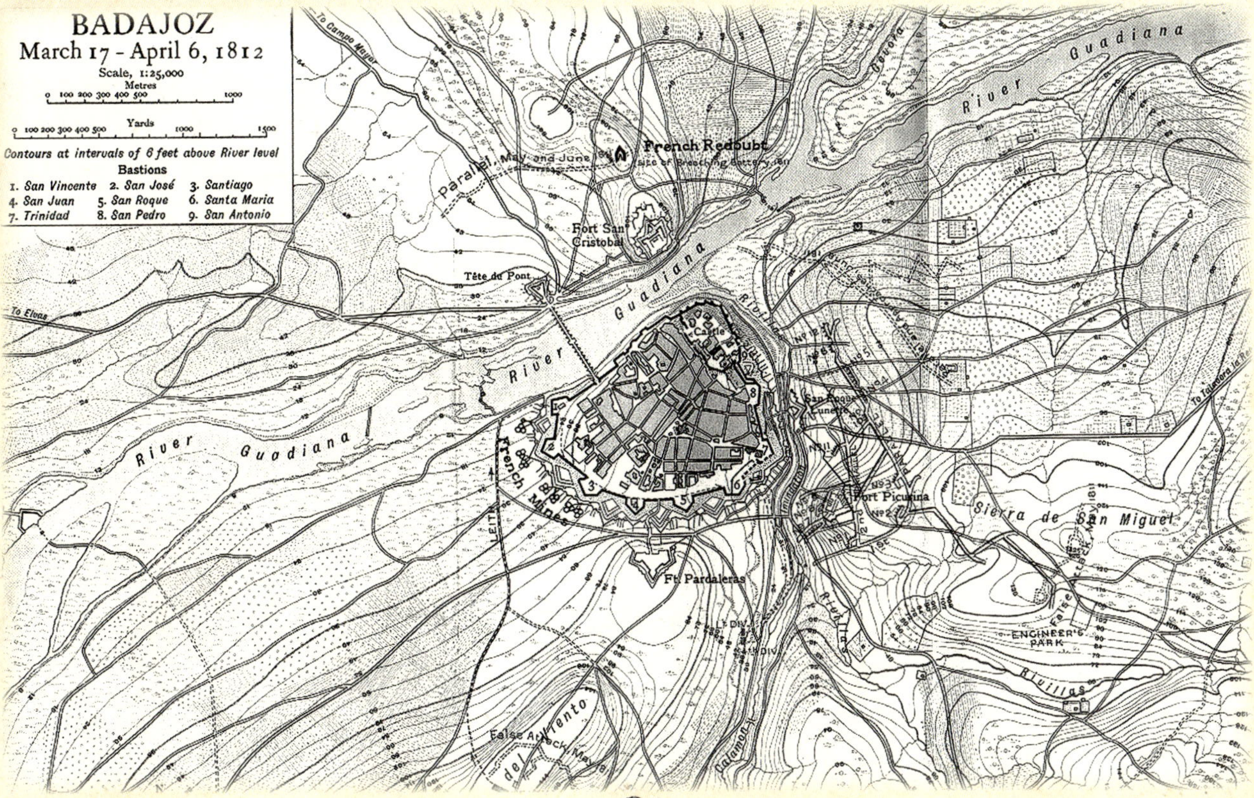

Blowing Up Bridges

Engineers can also blow up bridges. It takes two uninterrupted turns for them to lay the charges, which may or may not destroy the crossing. If the unit suffers four or more casualties or fights in close combat the engineers have to start the process again from the beginning.

At the end of the second turn roll 1D6 to see whether the charges go off successfully:

Wooden Bridges

D6 Score	Effect
1	The charges fail to go off and the engineers must start again.
2-3	The charges fail to go off, the engineers can try again next turn.
4-5	The bridge is set alight with one fire. An extra fire will start each turn and if not put out the bridge will therefore be destroyed in two turns.
6	The bridge is completely destroyed.

Stone Bridges

D6 Score	Effect
1	The charges fail to go off and the engineers must start again.
2-3	The charges fail to go off, the engineers can try again next turn.
4-5	The bridge is partly destroyed, only infantry units can still cross.*
6	The bridge is completely destroyed.

* The engineers must lay new charges. If the partly destroyed result is rolled again on the second attempt the bridge is completely destroyed.

Pontoon Trains

Pontoon trains were vast entities, containing enough men, horses, boats, wood and ropes to allow for the rapid construction of bridges. If your army contains at least three infantry divisions, it may include a single pontoon train at a cost of 25 points. The army may never include more than one pontoon train.

Movement & Fighting

Pontoon trains move as limbered foot artillery. They can only make standard moves, except on roads where they can march.

For fighting and morale purposes pontooners use the same rules as engineers.

Building Bridges

If your army deployment area contains a river you can build a pontoon bridge before the battle.

Bridges can also be built during the battle by a pontoon train which is stationary beside a river. It takes four uninterrupted turns to build the bridge and the pontoon unit must not be in any enemy unit's engagement range or suffer any more than four casualties. If either situation occurs the unit cannot carry on building.

Once a pontoon bridge is complete, all the rules for bridges apply.

British Pontoon train and escort.

Friendly Fire

Doctors will have more lives to answer for in the next world than even we generals.

— Napoleon Bonaparte

Throughout the Napoleonic Wars there were instances of confused troops firing on friends and allies. The fact that many armies dressed in a similar manner and were made up of numerous nationalities speaking different languages made it surprising that casualties from friendly fire were not more common. On still days the dense smoke clouds caused by black powder weapons added to the confusion of battle, as did mist, fog, heavy rain, snow and darkness. Troops could easily lose their bearings and mistake friend for foe, even firing volleys at them in the most extreme cases.

Troops who have received Stand and Fire!, Advance and Fire! or Retreat and Fire! as a random order, will fire a volley at an allied unit if no enemy troops are available as a target. This will only occur if the unfortunate target unit is a different nationality to those who are firing. A British unit, for example, cannot mistakenly fire on other British troops, but can accidentally target a Portuguese unit.

Normal rules for morale apply, so there is a good chance that any casualties sustained will affect the actions or morale of the friendly unit fired upon.

Movement Rates for Squares

Thus far, we have described squares as stationary formations. However, it was not uncommon for very highly trained or veteran troops formed in square to shuffle slowly over the battlefield. This was far too difficult for militia or even most line troops to achieve, so only elite and Guard infantry can perform this manoeuvre.

Elite or Guard infantry formed in a square within engagement range of any enemy can move up to 2" in any direction as long as they have not performed any previous movement or firing in the Tactical Orders phase; movement of the square is carried out in the Movement phase. The moving square cannot move closer to any enemy troops.

Elite and Guard squares can also move 2" in any direction as long as there are no enemy troops within their own engagement range. If the square moves into the engagement range of any enemy troops it must halt immediately.

Back to Back Fighting

Elite and Guard infantry units can form up back to back in order to fight on two sides. In order to do this the unit must remain stationary. When volley firing, or if involved in close combat, the unit can roll three dice for each direction. In the case of Guards, who receive an extra dice, the unit can use the additional roll in the direction of the player's choice. All normal rules for firing and combat apply. If the unit loses in close combat, or is forced to move, the unit moves as normal, facing away from the combat it has lost.

Surrendering

Any unit whose flank or rear comes into direct contact with enemy troops as the result of being forced backwards due to a lost command order or after losing a round of close combat must surrender immediately and is removed from the battlefield.

Any of your surrendering units count towards your percentage of infantry lost when counting to determine for victory or defeat; in the case of surrendering infantry or surrendering heavy cavalry, it is deemed that their colours are captured.

Prussian Landwehr Infantryman.

CLIMATE

It is easier to brave and threaten, than to conquer an enemy.

— Napoleon Bonaparte

Weather played an important role in the battles of the Napoleonic Wars. Wet conditions reduced the firepower of guns, while the extreme heat or severe cold encountered in conflicts stretching from the snow of Russia to the blazing sun of the Iberian Peninsula directly affected the fighting ability of troops.

Ground conditions had an impact on the effectiveness of artillery fire; on hard ground cannon balls bounced further, ploughing through greater numbers of troops, conversely they often became embedded in the wet soil of saturated battlefields before reaching their target.

Even mild windless weather played a part in reducing visibility by allowing fogs of musket and cannon smoke to linger on (and sometimes engulf) the battlefield.

These rules allow you to determine the type of weather on the day of your battle and the effect this will have on the fighting. If you are fighting a particular historical campaign then the weather conditions will be governed by those of the campaign itself, the details of which can be found in the relevant Army Lists.

If you are fighting a battle of your own creation then follow the steps below to determine the weather for that day.

First determine the season in which your battle occurs by rolling 1D6:

D6 Score	Season
1	Winter
2	Spring
3-4	Summer
5-6	Autumn

Ground Condition

Next, determine the prevailing ground condition for the season by rolling another 1D6:

	D6 Score to Determine Ground Condition			
Season	Snow	Soft	Normal	Hard
Winter	1	2-3	4	5-6
Spring	1	2-4	5	6
Summer	-	1	2-4	5-6
Autumn	-	1-2	3-5	6

French attack in the rain.

Prussian Reservist Infantryman.

Air Condition

Air conditions are determined according to the campaign from which your battle is taken:

A. Peninsular battles, invasions of France from Italy or the Pyrenees, Italian campaigns.

B. 1809 German Campaign

C. 1812 Invasion of Russia

D. 1812 Retreat from Moscow (only used as winter)

E. 1813 Wars of Liberation

F. 1814 Campaign for France

G. 1815 One Hundred Days Campaign

Roll 1D6 and consult the following chart for the current air condition.

If you are fighting during the spring or autumn add 1 to your D6 score. If you are fighting in summer add 2 to your D6 score.

D6 Score to Determine Air Condition

Campaign	Snow	Heavy Rain	Rain	Still	Clear	Hot
A	1	2	3	4	5-6	7-8
B	-	1	2-3	3-5	6-7	8
C	-	1	2	3-5	6	7-8
D	1-3	4	5	6-7	8	-
E	1	2	3-4	5	6-7	8
F	1-2	3	4	5	6-7	8
G	-	1-2	3	4-5	6-7	8

Many of the campaigns listed above only took place at a certain time of the year, so if you're recreating a historical battle or campaign, try to base your battle around the date of an historical engagement from that particular campaign.

When fighting a battle in the Hundred Days Campaign, for example, the fighting will occur in the summer – the Battle of Waterloo took place during this campaign on the 18th of June.

Ground Condition

Ground conditions primarily affect artillery firing ball shot and units moving at faster speeds.

The effectiveness of artillery firing ball shot onto unit columns, or lines if firing into flanks, can be increased or decreased depending on ground conditions. Use the relevant chart below to determine the effect (for normal ground use the chart in the Basic Rules):

Unit casualties inflicted when hit on soft ground, snow or if firing into marshland

Heavy Artillery	2 casualties caused on all columns (lines from the flank)
Field artillery	1 casualty caused on all columns (lines from the flank)
Horse Artillery	1 casualty caused on all columns (lines from the flank)
	1 casualty on line from the front or on artillery

Unit casualties inflicted when hit on hard ground

Heavy Artillery	3 casualties caused on all columns (lines from the flank)
Field artillery	2 casualties caused on all columns (lines from the flank)
Horse Artillery	2 casualties caused on all columns (lines from the flank)
	1 casualty on line from the front or on artillery

Movement on Soft Ground

Movement on soft ground is more difficult for units who decide to, or who are forced to, run. Roll as normal, but any scores of 1 or 2 cause a casualty.

Cavalry making an uncontrolled breakthrough on soft ground also suffer casualties on a score of 1 or 2.

JOACHIM MURAT, 1771-1815.
Marshal and King of Naples. He served in Egypt and Italy and, as a cavalry commander, took part in Napoleon's greatest battles.

CHARLES OUDINOT, 1767-1847.
Marshal and Duke of Reggio. Famous as commander of grenadiers. Fought at Austerlitz, Wagram, etc., but joined the Bourbons in 1814.

LOUIS DAVOUT, 1770-1823.
Marshal, Duke of Auerstädt and Prince of Eckmühl. Gained distinction in Egypt, at Austerlitz, Auerstadt, Eckmühl and Wagram.

MICHEL NEY, 1769-1815.
Marshal, Duke of Elchingen, Prince of Moscow. Served at Jena, Eylau, through the Russian campaign and at Waterloo.

Bavarian Infantryman.

Movement on Snow

Standard movement is unaffected by snow, but all marching and charging troops must test for fatigue casualties, roll 1D6:

D6 Score	Effect of Snow
1	The unit suffers 1 fatigue casualty and loses 3" of movement
2-5	The unit loses D3" of movement.
6	No effect

Movement on Hard Ground

Hard ground allows artillery to force march or run anywhere on the battlefield, all other normal rules apply.

ATMOSPHERIC CONDITION

Falling snow and rain reduce the effectiveness of muskets. In these conditions volley fire is only permitted at short and effective range, except for each unit's first volley of the game when it can fire as normal.

Artillery cannot fire at extreme range.

Heavy rain renders musketry almost useless, with the exception of the first volley, which can be fired at short and effective range only; subsequent musket volleys can only be fired at short range.

Artillery can only fire at effective and canister range.

The above rules for air condition also apply to battles in **fog or mist**.

Calm days caused a build up of smoke thereby making repeated fire difficult. In these conditions, once a unit has fired, place an extra smoke marker next to it (in addition to its first fire marker). This also applies in the case of Artillery Batteries.

The following turn the unit can only fire if on Stand and Fire! orders while being charged. At the end of the following turn, remove the smoke marker.

Hot days create a different set of problems for commanders, the main one being fatigue.

All movement on hot days, except for standard movement, requires a test for fatigue casualties.

Any units wishing to make a run move suffer a fatigue casualty on a 1D6 roll of 1 or 2.

Cavalry units making an uncontrolled breakthrough and any units that fail a charge, suffer 1D6 fatigue casualties.

In harsh ground conditions, armies can struggle to perform to their normal standards.

Austrian Infantry Officer.

Remember, gentlemen, what a Roman emperor said.
The corpse of an enemy always smells sweet.

— Napoleon Bonaparte

French General (Knötel).

PAINTING & COLLECTING

Painting a French Napoleonic Grenadier of The Old Guard

By Kevin Dallimore

The Grenadier of the French Imperial Guard (or Old Guard) is my idea of the archetypal Napoleonic infantryman, an impression that stretches back to the Airfix models I had as a child. Their blue uniform with vivid splashes of white and red is instantly recognisable and rewarding to paint. For a Napoleonic uniform it is also relatively simple featuring few colours and without too much lace, and it remained relatively unchanged throughout the Napoleonic era, so the versatile Guards can accompany most French armies right up to the end of the period.

This particular model is one of the Foundry classics, an imposing figure with his big bearskin and bristling moustache. He is depicted in 'full dress', which at the time was meant to be reserved for parade but was worn into battle by the Guard whenever possible. The only concession I have made to battle dress is to give him black marching gaiters in place of the white parade version.

Picture 1. Preparation.

Most metal models require at least some preparation or cleaning up before you can paint them. This involves the removal of excess metal such as mould lines and flash (wafers or spurs of metal that occur at the edges of the model) which result from the casting process. Use a good sharp knife for this; I use a scalpel by scraping the blade carefully along the mould line to carve it away. Be very careful with the knife to avoid painting your model with genuine blood red! Finish off by gently tidying and smoothing the area with a fine file. There is very little preparation required for this model as his simple shape makes him easy to cast. I usually attach the model to a piece of card to allow for safe handling during the painting process. I have not shown this in the pictures, but a piece about 20mm square will do fine.

1.

Picture 2. Undercoating the model.

I employ a three stage painting system working from dark to light on a basic black undercoat. Undercoat the model either by brushing on matt black paint or by using a black aerosol spray paint. I brushed on the undercoat for this model using matt black enamel, which forms a good basis to work from. Make sure the undercoat is not so thick that it softens the fine details of the model, or so thin that it does not give complete coverage. Once you have covered the whole model in undercoat put it aside to dry. It is worth checking back after about fifteen minutes just to make sure the undercoat has not pooled anywhere in big blobs. On this model the undercoat tends to collect between the legs. If this happens, brush the paint out over the model with a clean brush. You can also check for any misses in the undercoat at this point and paint them in.

2.

Picture 3. Painting the eyes.

Paint the eyes first. It is easier to do them now than to add them in once you have painted the rest of the face. Painting eyes can be a bit tricky to start with, so if you have just started painting and do not feel quite ready for such fiddly detail you can skip this part. Try to keep the eyes small, as large eyes make models look permanently surprised! Paint the whites of the eyes using **WHITE 33C** and then dot in the irises with **FRENCH BLUE 65B** or **DUSKY FLESH SHADE 6A** as on this model. If you make a mistake just black the offending eye out and have another go. If you make the eyes too big, tidy up around them with black paint until you are satisfied with the effect.

3.

Picture 4 to 6. Painting the flesh.

Paint the shade flesh colour, **FLESH SHADE 5A**, over the face (avoiding the eyes that you have just painted in). Try to leave just a very small amount of black showing where the flesh meets the bearskin and collar. Next paint the main flesh colour, **FLESH 5B**, over the shade colour leaving the creases of the face showing in the darker tone. If you accidentally paint over the creases you can always paint them back in again with the shade colour. Creases usually occur around the eyes/eyelids, below the cheek bones, around the mouth, the ears and the nose. The nose, for example, can be delineated by the main flesh tone with just a stripe down the middle and two dots at the sides above the nostrils. The final stage for the flesh is to apply the light flesh colour, **FLESH LIGHT 5A**. This goes over the first

two coats to form a highlight on the raised areas of the face, such as the nose, brows and cheek bones. Leave the bottom lip in the shade colour or paint it with **TERRACOTTA LIGHT 37C** as shown here. Paint the model's hands using the same process and try to leave a thin line of black between the fingers. Study the close up pictures to get the feel for where to place the layers of colour.

PICTURE 7 TO 9. PAINTING THE WHITE.

There is a lot of white to paint on this model! Models with large areas of one colour can be a blessing and a curse. The curse is that it will take quite a while to paint all the areas with that one colour. The blessing is that once the colour is finished, half your work on the model is done! As with painting the flesh this is a three-stage process. It is worth spending some time working out all the bits that are going to be white before you put brush to model, otherwise you might have to go back later and fill in bits that you forgot. I forgot the white plume holder and had to go back and paint it in afterwards.

Austrian Officer.

Paint all the white areas with **ARCTIC GREY SHADE 33A** first. Again try to leave just a tiny bit of black around where the clothing meets belts, hands or other bits of clothing, even where two areas of white meet. The amount of black you leave is up to you, neatness is really the most important aspect. Paint over the shade colour with lighter grey, **ARCTIC GREY 33B**. Try to leave the folds and creases on the model's clothing in the shade colour. You can detail the cords on the bearskin by doing a series of little blobs to represent the twisted cord effect. If you go over any areas that should be shade, just paint them back in afterwards. You can afford to paint over most of the shade colour in the case of white.

For the final stage paint pure white, **WHITE 33C**, over the other colours on the high points of folds and any extremities or ends of clothing to create a highlight. Again you can emphasise the cords of the bearskin. The broad white belts do present a bit of a challenge but can be done mostly white with little shade colour showing. I have a habit of making up folds and highlights where there are none sculpted on the model. If you try this approach be careful not to overdo it.

PICTURE 10 TO 12. PAINTING THE RED.

Paint all the areas that are going to be red with a dark red first, **BRIGHT RED SHADE 15A**. On this model the red areas are cuffs and turn-backs, plume, epaulettes and the piping on the collar and the back of the coat. Be careful not to get any red on the white areas and once again leave just a tiny bit of black around where the hands and other items of clothing meet. The only really tricky bit of red is the piping, but painting the piping now before the rest of the model's coat gives you the opportunity to easily correct and neaten it. The next stage is to paint over the shade colour with red proper. I deviated slightly from the standard three-colour system here by using **BRITISH REDCOAT 68B** as the next layer rather than **BRIGHT RED 15B**; this is because I wanted a more scarlet finish. Leave any creases in the

shade colour where you can find them. Red is one of those colours that can be a bit difficult to get to cover properly, so be careful not to over thin the paint before applying. If the paint does not cover well, wait until it dries and then apply a second coat. The final stage is to apply a highlight in a brighter colour to the high points and edges of the red areas, this time **BRITISH REDCOAT LIGHT 68C**. Some people prefer to leave red un-highlighted as it can make the red look orange if you overdo it, but you will not encounter that problem with the colours listed here.

Picture 13 to 15. Painting the Blue.

Start the dark blue of the uniform coat by painting **FRENCH BLUE SHADE 65A** all over the relevant areas. This will look very dark, but not every colour on a model needs to be bright. Then paint on **FRENCH BLUE 65B** leaving the folds and creases on the clothing showing through in the shade colour. Pay particular attention to being neat around the piping. The contrast of this layer will not look that marked with the shade colour, but it will increase when you varnish the model (see later). Finish off the raised areas with a blue highlight of **FRENCH BLUE LIGHT 65C**. The choice of which blue to use for the French is always controversial. I am not an expert on uniforms of the Napoleonic period so I chose the safe option of the standard **FRENCH BLUE 65**. Some would doubtless argue that this is not quite the right blue, but then have trouble describing exactly what the 'right' blue should be. Others will say my choice is too bright! I did think about using the more subdued **STORM BLUE 39**, which has a greyer tone to it, but I decided to save this for when I paint any of the dour Prussians.

Picture 16 to 18. Painting the Black.

The black areas; bearskin, gaiters, boots, cartridge case, scabbards, already have a shade coat of black thanks to the undercoat! Once again I have used a slight variation on the basic method here by painting over these areas with a 50/50 mix of **BLACK 34A** and **CHARCOAL BLACK 34B**, leaving the black undercoat showing in the folds and creases. Use a further layer of pure **CHARCOAL BLACK 34B** and highlight with a 50/50 mix of **CHARCOAL BLACK 34B** and **CHARCOAL BLACK LIGHT 34C** on the raised areas to finish off.

Pictures 19 to 22. Painting the Metalwork and the Rest of the Model.

Paint the shade colour, in this case **BURNISHED COPPER 43B**, over all the ironwork areas and the musket barrel. Try to leave just a little of the black undercoat visible where it meets other areas of the model such as flesh and clothing. Next paint over the shade colour on the ironwork with **CHAINMAIL 35A**, and then with a final highlight

of **SPEARPOINT 35C**, in each case leaving a small amount of the previous coat showing in the recesses and details. I chose to start with a bronze-like colour on this occasion to give the ironwork an older, distressed appearance as if it had been in use for some time.

The brass/gold bits, like the bands on the musket, use **BRAZEN 36A** as a base, then **SHINY 36C** and finishing with a bright highlight of **BURNING GOLD LIGHT 44C** for a really shiny look. You could enhance the antique look by letting a wash of **COPPER WASH 47A** run into the details of the metalwork.

22.

Picture 21 also shows the musket stock painted in the three stages of **SPEARSHAFT 13** and the hair/moustaches painted in the three stages of **BAY BROWN 42.** Picture 22 shows the model from behind where you can see the detail of the top of the bearskin, the Grenade painted on the coat's turn-back in the three stages of **ORANGE 3,** and the furry backpack painted in the three stages of **DUSKY FLESH 6**.

Picture 23, 24. Varnishing.

The model must then be gloss varnished. Make sure all the paint is thoroughly dry before commencing varnishing; it is best left to dry overnight. Be careful when applying the varnish, if it is too thick it may pool in the recesses of the model. When the gloss varnish has dried apply matt varnish. Be even more careful when painting on the matt varnish. It is much better to apply two thin coats than one overly thick one. Brush the varnish out well from any nooks and crannies where it is likely to collect and pool, just like when you were undercoating.

23.

24.

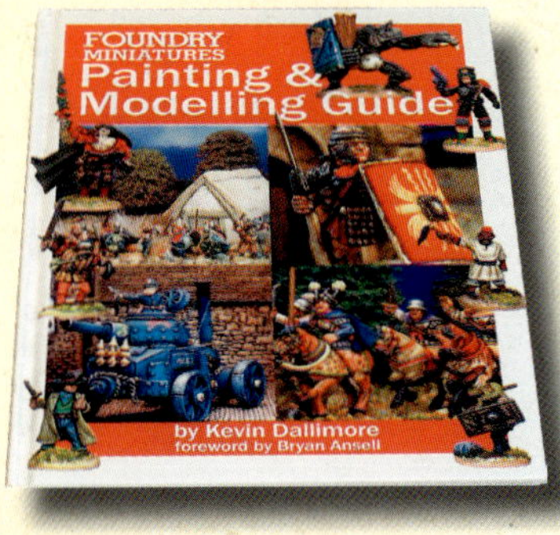

Kevin Dallimore's fully comprehensive guide to painting and building miniature models. Kevin describes every aspect of his celebrated painting style. By Foundry Publications

Picture 25. Doing a little bit less.

This Grenadier was painted by applying just two stages of colour to each area, mostly using the middle tone (the '**B**' paint) and the highlight (the '**A**' paint). The exception to this is the red, which used **BRIGHT RED SHADE 15A** and then went straight to **BRITISH REDCOAT LIGHT 68C** as a highlight. The two-stage method can save time when painting large numbers of identical models, but for an even faster approach I have also done a one-stage method.

25.

Picture 26. Doing quite a bit less.

This Grenadier Guardsman was painted by applying just one stage of colour to each area with no shading or highlighting. If done neatly this method of painting can still result in a very acceptable model although you have to make sure that the one coat is applied evenly and neatly; you will not be able to tidy it up with successive layers as in the three-stage method! Painting in this way does save quite a bit of time and is ideal for those needing to prepare an entire army for the tabletop. I generally used the middle tone (the '**B**' paint) for each of the colours except for the white (**33C**), which is the highlight of **33 ARCTIC GREY.**

26.

Picture 27 to 29. Doing that little bit extra.

This example shows the model painted using the same colour scheme as before, but this time I have used **four** stages of colour rather than three. The flesh, for example, has an initial shade coat of **TERRACOTTA 37B** before the standard three-stage flesh, giving it a richer, ruddier tone. I also made an effort to give him eyelids and painted a fleshy gap in his moustaches. The white here is done with **ARCTIC GREY 33C** as a base, then the three shades of **AUSTRIAN WHITE 67** and then a pure white highlight of **WHITE 33C**. The blue has a final extra fine highlight of a 50/50 mix of **SKY BLUE SHADE 21A** and **SKY BLUE 21B**. The black has a final fine highlight of pure **CHARCOAL BLACK LIGHT 34C** and the red has an extra highlight of a 50/50 mix of **BRITISH REDCOAT LIGHT 68C** and **ORANGE LIGHT 3C**. I ran a wash of **COPPER WASH 47A** around the detail of the plate on the bearskin. Also I could not resist doing one of these little chaps in white full parade gaiters, painted using the three stages of **CANVAS 8** with a final extra highlight of pure white, **WHITE 33C.** These extra touches help to emphasise the detail to an even greater degree and bring a little more life to the model.

27. 28. 29.

Picture 30 to 32. The finished model.

This picture shows how the finished model is enhanced by the addition of a scenic base. If you are using your models for gaming you will need a specific size of base conforming to the rules you are using. This Grenadier Guardsman is mounted on a Foundry pre-cut MDF base of 20mm x 20mm. If you're basing up models for

30.

31. 32.

Using a 50/50 mix of PVA glue and water, coat the whole base.

a unit, you may prefer to put four infantry models on the same 40mm x 40mm base.

Stick the model firmly to the base. I generally use a contact adhesive but any strong glue will do. When the glue has dried build up some 'ground' around the model's feet using some household filler or modelling putty to disguise

Dip the models' base in the sand.

the metal base, and then put this aside to harden. Using a 50/50 mix of PVA glue and water, coat the whole base (except for the edges) and dip the base in some sharp sand. The best sand to use is horticultural sharp sand from the garden centre as this is washed and cleaned of dust and salts, although it is a good idea to sieve out the larger lumps from the sand before use. Take the model out of the sand, shake off any excess and leave it to dry thoroughly. Be careful not to get too much glue or sand on the model itself!

French Grenadier.

Using a 50/50 mix of PVA glue and water to glue on the static grass in clumps.

Once the sand has dried, paint it **BASE SAND SHADE 10A**. Don't paint this on too thickly over the sand as it can clog up the grains; it's important that the surface texture of the sand should be clearly visible so that the next stages can highlight it effectively. Finish by dry-brushing the base with **BASE SAND 10B** followed by a final light dry-

Static flock applied with tweezers

brushing with **BASE SAND LIGHT 10C**. To add more visual interest to the base you can apply clumps of static grass (available from most hobby shops) with PVA glue as I have done here. Static grass is much more effective used in small scattered areas rather than covering the entire base. Allow the grass to stand for a couple of seconds before knocking the excess back into its container. Once the glue is dry the grass can be dry-brushed with **YELLOW OCHRE 4B** to tone down its shiny quality. (Picture 32 shows the left hand side of the Grenadier including detail of the tricolour cockade that covers the plume holder.)

Guards formed as a unit, with the more detailed painted models in the foreground.

I always use Foundry Paints as they fit in perfectly with my painting style, hardly surprising since I helped to create and develop them! For those of you that do not use Foundry Paints, the Foundry Colour Chart at the back of this book will help you match the colours referred to in the above text to your own choice of paints and hopefully help you produce a delightfully painted French Napoleonic Grenadier of the Old Guard that you can be proud of.

UNIT OF OLD GUARD

The picture above show the Guards formed as a unit, with the more detailed painted models in the foreground.

French Voltigeur.

PAINTING UNITS – A QUICK METHOD

Matt Fletcher: Whilst in an ideal world you always want to paint your models to the highest possible standard, sometimes you might want to get that unit of Cuirassiers, Grenadiers or Dragoons ready as quickly as possible, perhaps for a forthcoming event such as a tournament or perhaps because your regular opponent continues to crush you with ease and your army is in urgent need of reinforcements.

Having painted thousands of models over the years (seven hundred skeleton warriors in a week; four units of Napoleonic French infantry, each containing thirty-six models, in a week end), I have developed a number of techniques which allow me to paint units and armies to standard which looks more than acceptable on your tabletop, where models are being viewed from over a metre away. .

I normally paint for no more than one hour a day, five days a week, so in just five hours I paint about forty models a week or about two thousand a year – it takes me about eight minutes to paint each model. Using this method, tackling that pile of unpainted metal or plastic models which are causing your attic joists to creak is a real possibility.

Most of my painting involves painting units as part of an army so I will go through the process I use. The units selected are not only some of my favourite models but are also representative of the kind of rank-and-file troops anyone wanting to build an army will need to churn out in great numbers.

THE RULES OF SPEED PAINTING

I have a few rules of speed painting.

1. Keep it Bright!

I can't emphasise this enough: if you paint your models in dull, dirty colours, they will be lost on the table top. Any detail you have painted will be lost and your army will resemble an army of mud monsters! Some people say that bases painted in a colour resembling espresso coffee looks realistic and enhances the effect - no it doesn't, don't do it! As already mentioned we are painting units to look good viewed from about one metre away, so those white cross straps on your Napoleonic British infantry want to be white, placed on red-coated models.

Prussian Cuirassier Trumpeter.

2. Keep Them Neat and Clean!

If we are going to paint models in minutes we need to use good quality bright and clean paints; the Foundry range is ideal as the paints have so much pigment thereby' allowing good solid coverage. Faces especially need to be kept free from any paint splashes as these always seem to show.

3. Work on as may models as can cope with.

This is where most of the timesaving comes in. When painting models to a display standard you probably work on perhaps up to half a dozen models at a time, or perhaps as few as just one but when you want to paint units quickly, paint at least one whole unit at time, (more if you feel brave) but certainly at the least thirty models. One of the reasons you can save so much time is the same reason factories mass produce products, it's the economy of scale, in the end all forty-eight Austrian Fusiliers have black hats so paint them all in one go.

It is the planning of your time that is crucial and working on different units in sequence has advantages, as most of the early stages consist of filling in colour and therefore require the longest time to dry. So whilst one unit is half painted, start the next; each painting session paint one colour. This means, of course, that you are in a situation whereby you are constantly finishing models - which you'll appreciate has great motivational value.

4. The Work Area

You will need a clean surface to work on. I buy sheets of white faced hardboard and get the wood yard to cut them into 60cm squares – replace them when they get really paint splattered. The boards provide a place you can leave your models so you will able to get straight on with painting when you return to them. I keep mine in my utility room; this gives me the opportunity to apply a sneaky flesh wash when I'm meant to be doing the washing up!

I store my paints in a large wooden box, putting on my board only the paints and brushes I think I'm going to require (well that's the plan), but invariably you need that extra colour for something you didn't foresee. I find it almost impossible to paint on a cluttered, chaotic board!

Your work area is very important, I actually paint with great care with my board on top of the dining room table. This I find is a good compromise – you can still pretend to be joining in with your partner's conversation while getting on with painting; an occasional yes or no may be required though. This gives me more opportunity to paint than when I used to hide away in a room on my own. Or if looking after children you can watch them, with the noise of thousands of aliens being blasted or that cartoon you've already seen fifteen times in the background!

6. Planning Your Painting Time

This speaks for itself really – planning will help you get the most out of the time you have for painting. If you're going to be painting for short periods regularly (say an hour a night) then you don't really need to worry about drying time – you can just paint one stage in each of your sessions. If you're painting in longer, but less frequent sessions (for instance, a whole day on a weekend or on your day off) then you'll want to maximise your time by making sure you don't get left waiting for models to dry – consider painting a greater number of models, or having a variety of units on the go, (all at different stages in the process) so that the amount of time they spend drying will vary, thereby ensuring you have something to paint throughout your time.

7. Painting in them in order

I always lay my models out in lines, working always from left to right and top to bottom; in this way I know the order in which I painted the models.

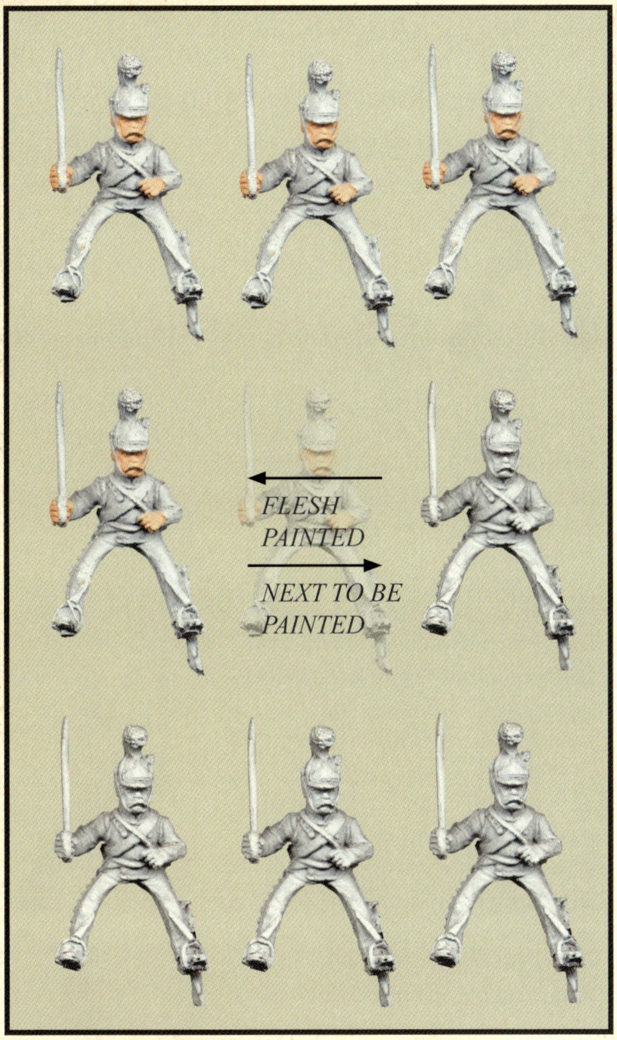

UNDERCOATING

Example 1, Painting a Unit of Austrian Napoleonic cuirassiers.

Cuirassiers are heavy cavalry and are therefore fielded in units of sixteen. In terms of models, the unit consists of thirteen troopers, two officers and a bugler, along with their sixteen mounts.

My main reference for the uniform was the Osprey *Men At Arms* series, Number 181. The text is written by the legendary Napoleonic Historian Philip Haythornthwaite, with colour plates by Bryan Fosten.

In this period Austrian cavalry units are distinguished by their facing colours, (the colours on their collar and cuffs, etc). After 1801 the Austrian army consisted of eight regiments of Cuirassiers and these were the heaviest and only armoured regiments in the army. The regiments had facing colours of dark red, black, grass green, light blue, dark blue and scarlet respectively. For no other reason than I fancied painting the dark red facings, I chose the 3rd Feldmarschall Herzog Regiment.

STAGE 1. PREPARING THE MODEL

The first stage of painting the unit is to clip of any excess metal in the form of mould lines and spurs or flash, as normal. Cavalry models often come in two forms: those which have the saddle and shabraque cast on the rider model and a bareback horse or those which come with all the tack upon the horse. Rider models often have a small metal bar between his legs and sometimes his scabbard; these are often put on to help with the casting process. The Foundry models come with the shabraque upon the horse. Neither style is necessarily easier to paint than the other, but it is worth bearing in mind which kind you are working with – you don't need to paint anything which is going to be covered up when the rider is stuck on his horse!

The horses may need extra cleaning between the legs, as excess metal often gathers there. You may wish to clip off some of slugs around the hooves of the mounts; I tend not to do this as it greatly reduces the strength of the model.

STAGE 2. UNDERCOATING THE MODEL.

There are a number of schools of thoughts as to which undercoat colour works best. Black gives shading; grey a great undercoat for highlighting for the coats; and white gives a clean and clear canvas to work from. Of course there is not a right or wrong choice. In this instance I have gone for white – remember those rules: rule 1, *keep it bright* and rule 2, *keep it clean.*

The men have been undercoated with white spay primer (each side having two light coats). You could of course paint them with a couple of coats of thinned down white paint (WHITE 33C).

British Hussar.

The horses are little more complicated, as I wish to paint them with mixture of colours. This is actually an area in which we as painters can add our own touch to the colour scheme by painting the majority of the horses in dark shades – blacks and bays – contrasting with the white of the troopers' coats. So, ten horses are sprayed or painted black, whilst six are sprayed or painted white. The trumpeter needs to be on a grey; we could undercoat his horse grey, but with spraying horses two colours, spraying one grey seems a bit over the top to me – choose either white or black, whichever you personally prefer to paint grey on top of.

STAGE 3. PAINTING THE RIDERS' FACES.

All of the riders have their faces painted in FLESH 5B. Try to be as neat as possible; you don't even have to paint into all of the nooks and crannies as we are going to give them a wash later on – remember the rules: keep your models clean! Just joking…

Anyway I like to paint the faces first; it always seems to bring the model to life!

You may need to thin your paint down by adding a little water. I never actually dip my brush directly into the paint pot; I simply shake the pot gently (make sure to check the lid is on securely first – a friend of mine forgot and ended up redecorating his ceiling!) then use the paint on the inside of the lid. A sable mix brush or any other suitable brush will suffice.

Don't worry about getting paint on the chinstraps or moustaches as these will need painting later, but make sure you paint the eyes. The officers are even quicker to do as they are wearing white gauntlets, so don't paint their hands. Lay them out in a logical manner if you want to do any more work on them today.

Once the paint on the face is dry, they should be given a FLESH WASH of COPPER LEATHER & METAL WASHES 47A thinned with a little water.

STAGE 4. BLACKING THEM UP.

The next stage is to paint all the black (BLACK 34A) on the models. This is the downside of white undercoating. I start with the boots, making sure to paint over stirrup irons and stirrup leathers so I give some depth to this area. The scabbard, cartridge box and breastplate (Austrian Cuirassiers only had breastplates in this period, which according to their French Adversaries put them at a considerable disadvantage!) are also painted black.

All of the helmet including its plume and chin straps need painting black. As I'm painting a lot of the model, I paint the bottom half in one painting session and the top half in the next, allowing for maximum drying time.

French Infantry Drummer.

STAGE 5. METAL WORK.

Starting once again at the bottom of the model, the stirrup irons, scabbard, water bottle, breast plate fittings on the cuirass and the sword, along with its hilt, are painted in SPEARPOINT LIGHT 35C.

STAGE. 6 PAINTING THE OVERALLS.

Next I paint the overalls in NIGHT SKY SHADE 62A being very careful around the stirrup leathers; a little visible black will only act as a shade. When dry, a quick highlight of NIGHT SKY 62B is applied. This is when I start to highlight the large areas of colour on the models.

STAGE 7. STIRRUP LEATHERS

This was the fastest stage yet! The leathers were painted in a BAY BROWN 42B.

STAGE 8. BUTTONS

The buttons on the grey overalls are dotted in with SPEARPOINT LIGHT 35C. Sometimes I paint them black first then paint over with the metallic. In this case the button is so fine and on dark colour, that I don't bother with the black and simply paint the metallic straight on.

STAGE 9. LET THERE BE WHITE!

One of the advantages of the white undercoat for models wearing a lot of white is the undercoat is slightly grey giving an instant shade. Starting with the scabbard straps, then the cross straps, then all of the remaining white areas, begin to paint in WHITE 33C, leaving a little of the undercoat in the deeper folds of cloth to act as shading. The collar, cuffs and turnbacks also require painting. The side of the helmet comb is painted white as we are going to finish it a yellow – even with modern paints, yellows and reds can still be a little tricky, and painting them first in white will help the paint to take.

The officer requires his gauntlets and the strap along the bottom of his armour painting white.

STAGE 10. FACINGS.

The facing colours were used to distinguish between each unit, originally coming from the coat lining in a contrasting colour to its outer one. This unit, as already mentioned, has dark red facings, so MADDER RED 60B is used. There is always a lot of debate about what is the correct shade or even colour; books often contradict each other and, worse, the printed colour often varies from the one originally painted by the artist. All a painter can do is take his best guess and ignore the picky criticism from those with nothing better to do!

The front of the collar, epaulette, cuffs and turnback are all painted.

STAGE 11. HAIR.

The models' hair and moustaches are painted with a mixture of different browns and; I avoid using black as I want a contrast from the models' black helmets. This example has BUFF LEATHER 7B; also used are DUSKY FLESH 6B, CONKER BROWN 54B and BAY BROWN 42B.

STAGE 12. GOLD.

The helmet plates and any buttons (this was a further way of identifying regiments, by using either gold or silver ones) on the coat are all painted BURNING GOLD 44B. The officer also has his sash and the edges of his cuirass painted. The trumpeter has his instrument painted gold (you could use brass, but that would involve finding and using another colour, and for speed the simple use of gold is much preferable).

STAGE 13. YELLOW.

The sword knot and the comb on top of the helmet, which was painted white earlier, is now painted YELLOW 2B. The officer's sash has horizontal flecks of yellow added to his gold sash; the trumpeter has his trumpet cords painted and he also has his plume painted yellow since this will help deepen the colour when the red is applied later.

STAGE 15. LEAVES.

Next paint the BRIGHT GREEN 26B leaves on the helmet.

STAGE 16. TOUCHING UP AND HIGHLIGHTS.

The final part of the process is a final touch up of any colours that have splashed onto other areas, or touching up any chips in the paint. Black and white are the most likely to require this. FLESH 5B is used to pick out the raised parts of the face, the cheekbones, chin and nose and the hands. Highlight the overalls with NIGHT SKY LIGHT 62C.

The trumpeter also has black flecks added to his cords and his yellow comb is gone over with BRITISH REDCOAT 68B.

The riders are now completed so its time to move onto the Horses!

Prussian Lancers painted by the Author.

Horses

In time of revolution, with perseverance and courage, a soldier should think nothing impossible.
— Napoleon Bonaparte

Cuirassiers rode dark-coloured horses; in some units, all riders were even mounted on black horses, though whether this was ever actually achieved in practice is debateable.

I believe that most collectors of models let themselves down buy simply painting their horses brown. Instead of painting them in the abundant number of shades and points horses actually possess. My advice is to go out and have a look at them, or better still go out and have a go riding one! (thought it must be said this will do little to improve your painting, and the two should not be combined).

The Foundry Painting Guide details lots of different horse colours to emulate. Also remember to use lots of contrasting colours on the horse in the relevant places; white socks and feathers look great on dark horses.

I will go through the process for each of the colours we are going to use one at a time; I will deal with the tack and furniture later.

Painting Black Horses

Stage 1.

Even if you have undercoated the horses with a black spray, it's still advisable to paint them with a slightly thinned down BLACK 34A. These horses have been hand undercoated, so I have not painted the shabreque yet.

Stage 2.

Both front legs, the near hind leg and a small blaze (white strip on the nose) are painted AUSTRIAN WHITE SHADE 67A. These are then gone over with WHITE 33C.

Painting Bay Horses

Stage 1.

Is the same as painting the black horses.

Stage 2.

The body, head and legs down to the knees are all painted in BAY BROWN SHADE 42A. Try to avoid painting over the tack, mane or the shabreque. Once dry a coat of BAY BROWN 42B is applied. You can use this technique with all the darker horse shades; CONKER BROWN 54B is good for liver chestnuts.

The Trumpeters Grey Horse

Trumpeters often rode grey horses to make them more distinctive, making them more recognisable when playing their instruments which were used to aid squadron and regimental officers to give out orders.

Stage 1.

The process is exactly the same as before: NIGHT SKY SHADE 62A, NIGHT SKY 62B and NIGHT SKY LIGHT 62C are used as a the main body colour.

Stage 2.

Once dry the mane, tail and legs are *dry brushed* in WHITE 33C; a white stripe is applied to the face of the steed.

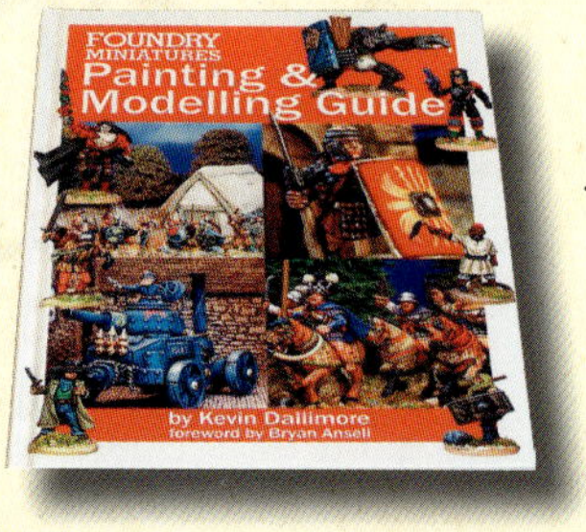

Kevin Dallimore's fully comprehensive guide to painting and building miniature models. Also by Foundry Publications.

French Infantryman.

Chestnut Horse

Unlike darker colour animals, where the hair on the mane, tail and legs are a darker shade, on chestnuts these are a lighter shade. CHESTNUT 53B is used. As this is a lighter shade, use a white undercoat – this will really brighten up the colour of the horse. When dry, the hair is *drybrushed* with BASE SAND SHADE 10A.

Painting the Tack

All the tack is now painted BLACK 34A – this includes all the bridal, reins, martingale (the strap around the chest joining to the bridal), girth (strap which goes under the stomach of the horse) and crupper (the strap which goes under the tail to the back of the saddle).

Once dry all metal fittings and the edge of the bit is painted with BURNING GOLD 44B.

> *I have destroyed the enemy merely by marches.*
> — Napoleon Bonaparte

Painting the Shabraque

I have left the rest of the horse black to demonstrate the technique more clearly. The whole area is undercoated in WHITE 33C. If you are painting over a black undercoat you may need to build up a series of layers. This is of course why I did not undercoat this part earlier.

The shabraque is then painted YELLOW 2B (this also thickens up the red). Once dry BRITISH REDCOAT 68B is applied, leaving about 2mm around the edge in yellow – this is a simplification of the real shabraque (with a black stripe at the very edge), but it helps keep it clean.

I paint all of one side, then when it is completely dry paint the other side.

Once painted, models can be based and varnished as you've already seen in the painting guide on page 98.

Russian Cuirassiers painted by the Author.

French Hussar.

ARMIES

French Cavalry attack the British Squares.

Massed British and French Cavalry come to blows.

Prussian General.

British and French armies close to combat.

Franco-Bavarian Army advances on the Russians defending the town.

Portuguese Officer.

Showcase

An Emperor confides in national soldiers, not in mercenaries.

— Napoleon Bonaparte

French Young Guard Officers

French Young Guard.

Much shedding of blood, many great actions, and triumphs,
toil and perseverance are the end of all things human.

— Napoleon Bonaparte

French Dragoon (Knötel).

THE NAPOLEONIC WARS

You do not get peace by shouting: Peace. Peace is a meaningless word; what we need is a glorious peace.
— Napoleon Bonaparte

In this section we present a history of the Napoleonic Wars, an overview of the armies and nations which fought them and a set of army lists allowing you to choose the forces for your own battles. This section is divided into a number of parts which largely fall into the following three categories:

THE CAMPAIGNS OF NAPOLEON

Spread throughout this section you'll find historical accounts of the most famous, pivotal and dramatic campaigns of Napoleon's reign and career. The campaigns described are: The Peninsular War (Napoleon's invasion of Spain and Portugal, and subsequent battles with the British), The German Campaign (Napoleon's attack on Prussia and the Austro-Hungarian Empire), The Patriotic War (Napoleon's invasion of Russia), The War of Liberation (Europe's revolt against Napoleon's rule following his retreat from Russia) and finally the War for France itself.

THE ARMIES OF THE NAPOLEONIC WARS

Each of these sections presents an overview of one of the major participants in the Napoleonic Wars, either the armies of Napoleon's France and her allies or those of his many enemies: the Austrians, Prussians, Russians and British. Each such section includes a brief description of the army, its troops, notable regiments, uniforms and colours. Also included is a painting guide and miniature gallery, an ideal guide to collecting an army. To avoid any confusion, however, these sections never contain any rules or game information – this can all be found in the relevant army list.

ARMY LISTS

Spread throughout this section are numerous army lists, each based upon a different historical army, representing the forces of various nations from the many different campaigns of the Napoleonic Wars. Army lists can be used to choose your forces for a battle and, by selecting troops up to an agreed points total, you can ensure a fair match-up with your opponent. The Army Lists for a particular period can be found after the historical information on the relevant campaign, and after the overview for each of the participating nations. Several of the nations were involved in more than one campaign, and were active over a number of years, so multiple army lists are included. An introduction to using army lists can be found on page 115.

Napoleon accepts the surrender of the Austrians.

French Grenadier.

The Revolutionary Wars and the Rise of Napoleon

In twenty five years, between 1790 and 1815 the military, political and social face of Europe changed completely. It was a time in which great men and great armies strode across the globe laying the foundations for much of the modern world. Pre-eminent among those who brought so much change was Napoleon Bonaparte.

In military terms the period marked the high-point of pre-industrial war, when strategy was still determined by the speed at which a man could march or a horse could gallop. Most battlefield tactics were still based on weapons that had dominated battlefields for almost a hundred years, and which dictated the manner in which armies had to be drawn up to gain maximum effect. But it was also the dawn of a new military era when new innovations were being introduced by the larger national armies; the use of rifles and skirmishing tactics on the battlefield and the use of rockets and, to a lesser extent, balloons were used as a precursor to modern aerial warfare.

The wars of the period, for this was not one continuous period of war but a series of wars interrupted by times of uneasy peace and alliances, stretched across the globe and involved armies of all the European states, and even America. For the historian and wargamer it offers endless opportunities for scenarios, pitting armies against adversaries based on real skirmishes, battles or campaigns, or exploring the 'what ifs' had fate determined a different course of events.

The French Revolution

The starting point can be taken as the French Revolution of 1789 when the Paris mob, faced with rising food prices, seized political control from the King. France underwent a social revolution watched with interest and then horror by

French Officer.

French army of the Revolutionary period.

the other nations of Europe, who initially welcomed the liberalisation of France but soon came to fear the spread of revolution to their own countries. It was not until 1791, two years after the revolution, however, that Prussia (modern-day Germany) and Austria indicated their wish to restore Louis XVI to power. The French National Assembly responded by declaring war on the Austrian Emperor in April 1792.

The French army had by now been decimated by the loss of most of its aristocratic officers, but the core of the ex-Royal regiments remained. These were supported by large numbers of volunteers who made up in patriotic fervour what the army lacked in experience and material.

Initial successes went to the Austro-Prussian forces under the Duke of Brunswick, but his army was held at the battle of Valmy, when the French General Dumouriez used his artillery to great effect and Brunswick was forced to withdraw. The tide of war now swung to the French. Louis XVI was stripped of what little powers he still held and was eventually executed in January 1793. The French armies under General Custine invaded Germany while Dumouriez marched into the Netherlands, seizing Brussels.

Britain and Spain now joined the war and French fortunes went into reverse. Custine was guillotined for his failures, 'to encourage the others', and the naval port of Toulon was occupied. The French government took the unheard of step of calling a 'Levee en masse', which allowed for the conscription of the whole male population. In the north, the allied armies in Flanders, including the British under the Duke of York, were pushed back, while in the south Toulon was besieged and recaptured, in part due to the efforts of a young artillery officer, Napoleon Bonaparte.

French Drummer.

The war continued to go well for the French with further successes in Germany and the Netherlands. By mid-1795 only Austria and Britain were still at war with her; the others had been either defeated or sued for peace.

The Rise of Napoleon

1796 saw the now General Bonaparte, who had skilfully elevated himself through the ranks by careful association with all the right political circles in Paris, given command of the French Army of Italy. With his ragged army he

immediately went on the attack, splitting the Piedmontese and Austrian armies and defeating each separately. Piedmont soon sued for peace. Pushing the Austrians back, Bonaparte successfully split and defeated the enemy forces in a whirlwind campaign. Much of the Austrian effort was aimed at relieving Mantua but Napoleon continued to defeat each Austrian force sent to relieve it, gaining, amongst others, the hard-fought victories of Arcola and Rivoli.

Eventually faced with French armies heading for Vienna from both Germany and Italy, the Austrian Emperor sued for peace. France acquired large areas of Italy and the Netherlands in compensation. Bonaparte was a national hero, but also a threat to the government. He was sent to Egypt on a mission aimed at threatening the growing British control of India. Although his army was cut off when the French fleet was destroyed at the Battle of the Nile by Nelson, he continued to add to his success with several crushing victories over the local rulers, the Mamelukes.

In 1799 a new Austro-British alliance was formed that initially saw success in both Germany and Italy. Bonaparte, sensing he was sidelined, left his army and seized control of France. In an audacious move he led his army over the Alps and back into Italy which had been all but lost. Losing touch with the Austrian positions he stumbled into their main army at Marengo where, with only eighteen thousand men, he faced a force almost twice the size. The battle was almost lost but the late arrival of General Desaix saved the day, although it was to cost Desaix his life. The Austrian army was decimated and both Austria and Britain sued for peace. Napoleon and France were now the dominant force in Europe and in 1804 Napoleon was proclaimed Emperor.

The Early Napoleonic Wars

The peace arranged with Britain was to be short-lived and in early 1803 Britain declared war. Napoleon responded by stationing his troops at Boulogne in preparation for an invasion. Austria and Russia sided with Britain, and in 1805, with his invasion plans shattered by the loss of his fleet at Trafalgar, Napoleon turned his army south to face the growing threat from these new enemies. In one of his finest campaigns he first surrounded and then captured the Austrian forces under General Mack at Ulm before pushing on to Vienna. The remaining Austrian forces fell back and joined up with the Russian army which had been marching west to support them, but in the face of Napoleon the combined armies retreated east. To force a battle Napoleon moved his army onwards tempting his foes with his apparent vulnerability. The Emperors of Austria and Russia took the bait. At Austerlitz in December 1805 Napoleon withdrew from the Platzen Heights and encouraged the Allied forces to attack his right wing. As they did, his main force stormed the Heights and decimated the Austrian and Russian forces which fled the field in disorder. It was a tactical masterstroke and perhaps Napoleon's greatest triumph. Austria again sued for peace.

However, Britain and Russia remained in the war and were now joined by Prussia. The Prussian army, which had been all conquering under Frederick the Great, was still stuck in its old ways despite trying to modernise. In October 1806 Napoleon concentrated his forces in Bavaria and by a rapid thrust took the Prussians by surprise and threatened Berlin. On 14th October he engaged a Prussian force at Jena, believing it to be the main army, unaware that at the same time Davoust was fighting the main force at Auerstadt fifteen miles to the north. The combined victories destroyed the Prussian army and to all intents and purposes it could take no further part in the war.

Napoleon now turned to the threat from Russia and invaded Poland. In February 1807, in the midst of a blizzard, the two armies fought out an indecisive battle at Eylau. Both armies suffered terribly but it was the Russians who withdrew. They resumed the offensive in the spring but were held by Napoleon at Friedland who then proceeded to take the Russian army apart, it being caught with the River Alle to its rear with no means of escape. On 19th June Napoleon reached Tilsit where the Russian Emperor, Alexander, sued for peace.

Napoleon could now claim to be master of Europe.

Napoleon had inherited a formidable army, a fighting machine created by the chaos of the Revolution that he had taken to another level. The French army could move rapidly unencumbered by supply lines, foraging instead from the land. The army was also divided into separate corps, each a small army in itself, able to fight and hold an enemy until support arrived. These factors, together with Napoleon's eye for strategy – largely based upon his often repeated ploy of splitting the enemy forces and defeating them in detail – and his ability to recognise and exploit the key moment in any battle, was to make the French victorious across Europe for over ten years.

Madame Guillotine

French patriot.

He that makes war without many mistakes has not made war very long.

— Napoleon Bonaparte

French Hussar (Knötel).

ARMY LISTS

Sometimes a single battle decides everything and sometimes, too, the slightest circumstance decides the issue of a battle. There is a moment in every battle at which the least manoeuvre is decisive and gives superiority, as one drop of water causes overflow.

— Napoleon Bonaparte

The Army Lists in this book are designed to allow players to choose armies to an agreed points value. In most battles opposing forces will start with an equal points value to create a balanced game but some scenarios, campaign games or historical re-fights might require armies of varying sizes. For instance, both players may agree to each choose an army to a total of 500 points, giving roughly equal forces who stand an equal chance in the battle.

The lists are not based on any specific corps or division but are designed to represent a typical army from each campaign. You can therefore use them either as an aid in recreating the army from a particular battle or simply select your own force from scratch.

Each army list is broken down into a number of sections, primarily **Special Rules**, **Divisions, Brigades & Batteries**, and **Units**. Each of these is explained below.

USING THE ARMY LISTS

SPECIAL RULES

As you'd expect, this part of the army list details any special rules the army is subject to – any ways in which they deviate from the normal game rules presented elsewhere, any special abilities unique unto themselves or any special bonuses, penalties or modifiers they may be subject to. Some special rules affect the army as a whole, others are specific to individual troop types within the army. Likewise, where a nation has more than one army list, some special rules will apply to all the armies belonging to that nation, others will apply only during specific periods. For instance, during the Peninsular War, any unit in the French army is able to adopt the special Mixed Column formation; during the War for France, only Old Guard units are able to do this.

Only the special rules attached to a particular army list apply to that army – just because the French army has a special rule in one campaign does not mean it applies to French armies fighting in others!

DIVISIONS, BRIGADES & BATTERIES

Listed at the start of each army list (sometimes after a short historical introduction) are the divisions, brigades and batteries available to you – in short, the different fighting bodies that will go to make up your army; infantry, cavalry, artillery, skirmishers and all the rest. Divisions, brigades and batteries are listed in charts like this one:

Type	Availability	Consists of:
Infantry Division	Your army may include any number of infantry divisions.	3-5 infantry units, plus a divisional general. Up to two units may be light infantry; up to three units may be elite (either light or line).

Russian Cuirassiers.

Type: This is the type (or simply the name) of the division, brigade or battery.

Availability: This column describes how many of each type of division, brigade or battery are available to your army and any requirements you have to meet in order to field them – you may have to field a certain number of cavalry brigades in order to include a battery of horse artillery, for example.

Consisting of: This explains which individual units (or guns in the case of artillery) make up the division, brigade or battery. Sometimes the number and type of these constituent units is fixed; other times you are able to choose within certain restrictions. Either way, the units are purchased from the unit list, at the cost shown there (see opposite).

UNITS

After the division, brigades and batteries, the individual units of the army are listed, in a chart like this:

Troop Type	Points Cost	Unit Size	Morale/ Class
Line Infantry	20 points per unit	6	Line Infantry

Troop Type: This is the type (or simply the name) of the unit in question. Very often, general, non-specific names are used – as is the case here. 'Line Infantry', and other general terms like it, invariably represent a variety of different specific units. In game terms, these units were all sufficiently alike in their training and equipment to be considered identical. When collecting your army, however, it's worth remembering that you have a choice of which miniatures you will actually use to represent each of these units. More information on this can be found in the overview of each army.

Points Cost: This is the cost to field the unit. The total cost of all the units in your army should not exceed the agreed points total for the battle. You must pay the points cost for every unit in the division (so a division of five Line Infantry units would cost 100 points). In the case of artillery, you are sometimes allowed to field a half battery: the cost of doing so is written after a slash, with the cost of a full battery before it, like this: 60/30.

Unit Size: This is the number of bases of miniatures which go to make up the unit. If you have used a different basing system for your miniatures (such as individually based figures, for instance), you should group them in such a way as to approximate the appropriate number of bases (see Unit Sizes, below, for more information).

Morale/Class: Morale and Class encapsulates a variety of information – principally the unit's morale class and its type in terms of movement, shooting and combat. Thus, line infantry roll on the Line Infantry chart for lost orders, and move as infantry; Guard cavalry roll on the Guard Cavalry chart for lost orders, move as cavalry and follow all the other rules for Guard units. Very occasionally, a unit may have a slightly confusing name which doesn't make it's class clear – in all cases it's the class and type shown in the Morale/Class column, not the Troop Type, which should be used.

GENERAL INFORMATION

As well as the specific information contained in each army list, the following general considerations apply to selecting your army and fielding units on the tabletop.

Unit Sizes

The recommended unit sizes are: six 40mm square bases with four miniatures on each for infantry (24 models); six bases of light cavalry or eight bases of heavy cavalry miniatures, mounted in pairs on 40mm x 50mm bases.

Skirmishers are mounted in pairs on 40mm square bases.

Artillery batteries vary in size; each list details not only the size of the battery, but also the ratio of guns and howitzers and each gun should be mounted with its crew on an 40mm x 50mm base. This is summarised below:

Unit Type	Unit Size	Miniatures per base	Base size
Infantry	6 bases	4 per base	40mm x 40mm
Light Cavalry	6 bases	2 per base	40mm x 50mm
Heavy Cavalry	8 bases	2 per base	40mm x 50mm
Skirmishers	Varies	2 per base (may vary)	40mm x 40mm
Artillery	Varies	1 gun plus crew	40mm x 80mm

Very occasionally (most notably in the case of Austrian armies) these unit sizes will vary somewhat. In any case, the size of each unit is noted in the army lists – the size shown there always takes precedence over anything you may have read elsewhere in these rules.

Unit Command

All units are assumed to include a variety of command – officers, standard bearers, musicians and so on. Any unit can include miniatures representing its command, generally one base per unit. Use of such a 'command base' is not essential, but can be very useful in denoting the unit's formation or morale state – the command base will always feature at the front of the unit, and can be turned around or thrown into disorder to indicate lost command, and so on.

French Infantry Officer.

THE PENINSULAR WAR

Men soon get tired of shedding their blood for the advantage of a few individuals, who think they amply reward the soldiers' perils with the treasures they amass.

— Napoleon Bonaparte

Napoleon was master of continental Europe but Britain was still undefeated. Although she was weak militarily, she was the paymaster of the other countries of Europe and as such could still undermine Napoleon's ambitions. Napoleon, unable to invade Britain, instead attacked her economic base. Through the Berlin Decrees he established his Continental System, forbidding any commerce with Britain. Spain and Portugal, however, remained outside his control leaving a gap through which British goods could still be imported.

To close this breach, General Junot was sent at the end of 1807 to occupy Lisbon, and in March 1808 Murat led another army into Spain. The Portuguese Royal Family fled to Brazil, while the Spanish Royal Family was compelled to renounce the throne. The Iberian Peninsula was under French control and Napoleon placed his brother Joseph on the now vacant throne of Spain.

Both the Portuguese and Spanish armies were, at this time, poorly led and organised and were no match for the French but, spurred on by the church, the Spanish people rose in revolt. The city of Saragossa successfully resisted French attempts to capture it and in July 1808 a French army under General Dupont, which had been ordered to Cadiz, was defeated by a Spanish army at Bailen – the Spanish army's sole success of the war.

Junot was effectively cut off in Portugal and in August faced a landing by British forces under Sir Arthur Wellesley, who had made his name as a commander in India. Wellesley took the offensive and marched on Lisbon, first defeating the French at Rolica and again four days later at Vimiero. Command was then handed over to Generals Dalrymple and Burrard who had just arrived from England and who proceeded to negotiate the removal of Junot and his forces from Portugal in British ships under the terms of the Convention of Cintra. When news of this reached Britain it was greeted with outrage and the commanders were immediately summoned home to answer for this arrangement. Dalrymple and Burrard were never to command again.

Command of the British army in Portugal now passed to Sir John Moore, the leading British commander and an exponent of light infantry tactics. He led his British army into Spain on the promise of support from the Spanish authorities but on finding himself unsupported and up against superior French forces under Napoleon himself (who had come to take personal command), Moore turned and led his army northwards to the ports of Corunna and Vigo during the winter of 1808-09. Facing terrible conditions the British army managed to stay ahead of the French (now led by Marshal Soult) and in a final act of defiance the British army turned at Corunna and defeated Soult's army thereby allowing time for the troops to be embarked onto waiting ships. All of this was however, at the cost of Moore's own life.

Portuguese infantry formed a major part of Wellington's army in Spain.

French Officers.

Britain still held Lisbon, and Soult now moved south. Wellesley was put back in command and in a bold move crossed the river Douro and defeated Soult at Oporto, before advancing into Spain. Promised support from the Spanish forces under General Cuesta, Wellington faced the French at Talavera to the south-west of Madrid. The promised support was not forthcoming but after a desperate fight Wellesley was victorious. Unable to count on the Spanish, however, he was forced to retire to Portugal. For his victory he was given the title of Lord Wellington.

For the next year, Wellington spent his time carefully reorganising the Portuguese army under British officers and also building defensive lines (Torres Vedras) outside Lisbon. In the summer of 1810, Massena attacked Portugal and Wellington carefully withdrew before defeating him at Busaco. Here the French, attacking in columns, came upon a typical Wellington strategy. The British and Portuguese forces drew up behind the sky-line and fought in line to gain maximum firepower. Despite this victory Wellington continued to fall back to Lisbon and once behind his prepared lines sat out the winter. Unable to advance, Massena's force faced starvation and eventually withdrew.

The war in 1811 centred on the frontier fortresses of Fuentes d'Onoro, Ciudad Rodrigo, Almaida and Badjoz, which together controlled the two main roads into and out of Portugal. An indecisive battle at Fuentes d'Onoro followed, as did the butchery at Albuera, where a British force under Marshal Beresford was almost defeated (but where the steadfast conduct of the British infantry eventually saved the day). Elsewhere the French consolidated their control over Spain.

In 1812 Wellington felt able to take to the offensive, capturing first Ciudad Rodrigo and then Badajoz in two bloody sieges before marching on Madrid, defeating Marmont at Salamanca. Wellington briefly held the capital, but after an unsuccessful assault on Burgos, was again obliged to retreat to Portugal.

1813 brought a turning point in the war. Driving north-west, Wellington delivered a crushing blow to the French at Vitoria, effectively ending any control they had over the country. Soult now took command of what remained of the French forces and attempted to defend the Franco-Spanish border. Wellington pushed him hard and in a series of actions forced the passes of the Pyrenees and took the border fortresses of San Sebastian and Pampelona. The French were now pushed out of Spain and in February 1814 they were defeated on French soil at Orthez. In April Wellington inflicted the final defeat on Soult at Toulouse. Ironically Napoleon had abdicated four days before.

Spain and Portugal during the Peninsular War 1808-1814.

FRENCH ARMIES

UNIFORMS AND COLOURS

FRENCH LINE INFANTRY

The mainstays of the French army were its line infantry, the fusiliers. Although such basic, musket-armed troops had been common to all European armies for centuries, the training and organisation introduced by Napoleon were revolutionary, and soon adopted by his enemies.

The enlisted men were known as 'fantassin' and their typical dress, after the adoption of the shako during the winter of 1806-7 consisted of gaiters (black in winter and white in summer), as well as the alternative use of the white uniform, introduced the previous year. The white uniform was reintroduced to the French army because of a shortage of indigo dye; traditionally Napoleon was supposed to have been so horrified by the sight of the white uniforms covered in blood on the field of Preussiches-Eylau in 1807 that he restored the blue uniforms. However, even if this were true, it took until 1809 to entirely replace the white (it also took until 1809 to get rid of all the bicornes).

Not that many units were actually supplied with the white coats, those that were, or might have been, together with their facing colours are as follows (units where this was possible but uncertain are indicated in *italics*):

Regiment(s)	Facings
3rd, 4th, 8th	Dark green.
12th, 13th, 14th, 15th, 16th	Black.
17th, 18th, 21st, 19th, 22nd,, 24th	Scarlet.
25th, 27th, 28th, 32nd	Capucine (an orangey-brown).
33rd, 34th, 36th	Violet.
46th	Sky Blue.
53rd	Pink.

Unit of early French line infantry, consisting exclusively of fusiliers and command.

Early French infantry.

Voltigeurs

Voltigeurs were the skirmishers and light companies of the French line infantry. After 1808, a company of voltigeurs was normally included alongside four companies of fusiliers and one company of grenadiers in each infantry unit. The voltigeurs could be detached to act as skirmishers.

The Voltigeurs were the most colourful element of a French infantry battalion with plumes of yellow, green, yellow-over-green, green-over-yellow, yellow-over-red, red-over-yellow, yellow-over-blue (3rd in 1812), blue-over-yellow (8th in 1812-13), or red-over-green (4th in 1809).

Unit of early line infantry in six company arrangement – four bases of fusiliers (one of which is command), one base of voltigeurs and one base of grenadiers.

Shakos (headgear) were often decorated with yellow top bands, and sometimes bottom bands as well. The shako cords were usually yellow or green but sometimes white (the 65th had green cords with red tassels in Spain 1810-14). Collars were normally yellow (chamois), even when the white uniform was in vogue (33rd wore violet in 1807), and were edged red or white. The most popular colour for epaulettes was green, with usually yellow, but sometimes red decoration.

French Voltigeurs.

Grenadiers

Grenadiers had once, as their name suggests, been grenade-armed troops, though by the time of the Napoleonic Wars grenades had declined in use and the name was instead retained for elite units, especially those skilful in an assault. Grenadiers were usually included alongside fusiliers and Voltigeurs in line infantry units, though on occasion were withdrawn from these formations and combined into units of their own. Some elite units of grenadiers were assembled on a permanent basis.

The colours for Grenadiers were somewhat more conservative with red used throughout (there are a few cases of white shako cords). Variation here occurs in the positioning of the colours with plain or coloured shako bands being the most notable option.

French Grenadiers.

Grenadiers in Bearskin

During the period in question the bearskin should have been replaced by the shako but many companies retained it. During the early campaigns of the Napoleonic period, as well as during the Peninsular War, the French developed the practice of taking the elite companies away from their parent units and forming 'elite' battalions. Interesting units can therefore be formed from mixtures of differently dressed grenadiers and voltigeurs.

French Grenadiers in Bearskin.

He who makes war for National independence must be enabled to count upon the union of all resources, all the wishes, and the concurrence of all the National authorities.

— Napoleon Bonaparte

Drummers

Before the introduction of the Imperial livery in the Bardin regulations of 1812, the colonels of the line infantry regiments were left to their own devices in regard to how they chose to dress their drummers and musicians. There was considerable competition in this field and vast amounts of money were spent until the Emperor attempted to curb some of the excesses in 1812.

French Drummers.

French Infantry in Greatcoats 1807-15

First introduced officially after the Austerlitz campaign of 1805, the capote (greatcoat) became an indispensable part of the French infantryman's kit. It was available in a variety of colours, grey was the official, dependent on the availability of cloth. Officers usually wore blue though they too might have to put up with whatever with was supplied.

French Voltigeurs and Grenadiers in Greatcoats.

French Infantry in Campaign Dress

With a cloth cover over his shako and trouser either over or replacing his breeches and gaiters, this is a representative example of what the French Fusilier at war looked like. The shako cover could be of a waterproofed material (usually black) or latterly, more commonly, of local cloth of any shade of brown, white, grey or even ticking. White was probably the colour of choice for trousers but, particularly in Spain, local homespun brown cloth was plentiful and cheap enough to do as a replacement.

French Fusiliers and Grenadiers in Campaign Dress.

French Grenadiers & Voltigeurs in Campaign Dress

Most of the comments above are relevant here, in addition the full dress plumes were replaced by tufted pompoms in red for Grenadiers, any two of green, yellow and red for the Voltigeurs.

French Grenadiers with Bearskin in Campaign Dress

The Grenadiers, in these circumstances, normally removed their cords, but unlike the Guards there is not much evidence for the use of waterproof covers. They may have stowed the bearskin in a bag of blue and white ticking and tied it to their knapsack on the march. On such occasions the forage cap 'bonnet de police' (see below) would have been worn.

Forage Caps

The forage cap, or 'bonnet de police' was an alternative hat worn by troops in campaign dress. It was usually blue with red piping for Fusiliers and Grenadiers and often with yellow piping for Voltigeurs. The front was usually decorated with grenades and hunting horns for the elite companies and sometimes carried a crown "N" for the centre companies. The bonnet de police remained blue even when the uniform was white.

French Infantry in Bonnet de Police.

FRENCH LIGHT INFANTRY

FRENCH CHASSEURS

In a line infantry battalion the men of the 'centre' companies were called fusiliers, the men of the light infantry company were called voltigeurs and the men of the 'heavy' company Grenadiers.

In the light infantry the voltigeurs were known by the same name but the fusiliers and grenadiers were called chasseurs and carabiniers respectively. Although initially used in a skirmishing role it is probable, by the height of the First Empire, that light infantry battalions were used in exactly the same way as their 'ligne' comrades.

Unit of early French light infantry with Chasseurs, Carabiniers and Voltigeurs.

This did not stop them maintaining those elite elements of dress and equipment which were part of the original light infantry ethos. The chasseurs for example had originally worn fringed epaulettes and carried a sabre, a distinction reserved for the elite companies of line infantry. Indeed they fought, with some success, to maintain these privileges even after 1812 when the epaulettes and 'sabre-briquet' were officially withdrawn, by the Bardin regulations (they were also taken away from the line and light voltigeurs).

French Chasseurs.

Chasseur distinctions usually included red-piped-white collar, green epaulettes (sometimes with red decoration) and company pompoms (plumes, when worn, were, officially, red-over-green, though in 1811 the 17th possessed white-over-green). Known variations in dress included green collars, white decorations on epaulettes and plain green plumes (cap cords were usually white or green).

All troops wear the type of short gaiter with a top styled to resemble a light cavalry boot. These were edged with lace and decorated with tassels; colours were officially green for chasseurs, red for carabiniers and yellow for voltigeurs. The most common type of sleeves for this period were square, in the same style as the line infantry, and whilst the body of the cuff was normally the blue of the coat the flaps were usually red for chasseurs and grenadiers and yellow for voltigeurs. Where pointed cuffs occurred they were most often blue or sometimes red.

French Chasseurs.

French Line Infantry in Camping Dress.

French Carabinier

Like their Grenadier counterparts in the line infantry the carabiniers tried to keep their bearskins, however some were reduced to shakos decorated with red cords and plume, with more red distinctions on epaulettes and collar. Variations included the 10th (plain shako bands with either red or white cords being recorded in 1809) and 8th (red top and bottom shako bands).

French Voltigeur

Voltigeurs took on the usual arrays of yellow, red and green. Combinations include:

Regiment(s)	Facings
7th (1809)	Yellow-over-red plume, yellow shako bands and cords, yellow collar, piped red; the epaulettes had yellow boards, red fringes and white decoration.
16th	Yellow-over-green plume, white cords, epaulettes with yellow boards and green fringes.
8th (1809)	Red-over-yellow plume, epaulettes yellow boards with red fringes.
27th (1809)	Yellow-over-green plume, green epaulettes with yellow decoration.
10th (1809)	White cords and piping to collar.

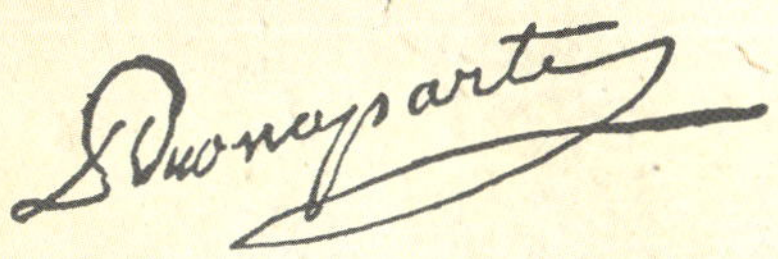

French carabineers.

French voltigeurs.

French Voltigeurs or Carabiniers in Full Dress and Colpack

A distinctive variation was the wearing of the colpack (fur hat) by the elite companies in the light infantry. Examples included the 17th (voltigeurs; yellow-over-green plume, white cords, red bag), 7th (1809 carabinier; red plume and bag, no cords), unknown (voltigeur; yellow-over-red plume, white cords, yellow bag) 27th (yellow bag). The colpack was also worn extensively by senior musicians and other members of the 'tete de colonne'.

French Carabineers in Colpack

Drummers

Drummers colours included: 10th (1809 all medium blue piped white or later with yellow facings), 27th (1809 turquoise faced maroon laced white), 8th (1809 blue faced green with tricolour lace), 17th (blue faced red with white lace), 16th (red faced green with white lace), 7th (blue faced red with white lace), 15th (1807 blue faced red with tricolour lace). Cornet players were usually dressed like the rest of the men with little in the way of distinctions.

French Voltigeurs and Carabiniers in Campaign Dress

On campaign plumes are again reduced to pompoms; red for carabiniers, yellow for voltigeurs.

French Voltigeurs and Carabiniers in Campaign Dress and Bearskin

Cords only present at the top of the bearskin to allow the contrasting white or red option to appear, all other comments as above.

French Mounted Infantry Officers

The surtout was an undress coat, single breasted with plain cuffs (sometimes piped red) and collar either plain (sometimes piped red) or with company distinctions. The tails were always long and turnbacks blue (sometimes piped red). The usual epaulettes from the 'habit' were worn. It was generally an officers coat but was adopted by some senior guard regiments, both horse and foot from 1806-7. Even during the period of the white coats the surtout was always blue, indeed the Otto manuscript of 1807 shows the widespread use of the surtout by officers in conjunction with the white coat for the lower ranks.

Once again most of the light models infantry will do very well for Italians and Neapolitans.

French Light Infantry in Campaign Dress

On campaign, the light infantry were often issued with blue trousers, something rare in the line. Waistcoats in both line and light were sleeved affairs, sometimes worn as a top coat, with or without epaulettes etc. being attached. They were white for the line but either blue or white in the light. Where officers are shown wearing their greatcoats, known as a 'capote', rolled over their shoulder, they would normally be coloured blue (whereas the soldiers version could be any variation on white, brown or grey).

French Light Infantry in Campaign Dress

French Mounted Infantry Officers.

French Infantry in Campaign Dress.

The Imperial Guard

If you wage war, do it energetically and with severity.
This is the only way to make it shorter and consequently less inhuman.

— Napoleon Bonaparte

Throughout Napoleon's reign, the men of the Imperial Guard were the very finest soldiers of the French army. The Imperial Guard comprised numerous regiments of both infantry and cavalry, and also included artillery. Generally, the Imperial Guard are divided into their groupings or periods known as the Old Guard, Middle Guard and Young Guard respectively. The Old Guard is the title given to those men who had served since the beginning or very early years of his reign, the Middle Guard was composed of those regiments raised during Napoleon's earliest and most successful campaigns of conquest (1805-1809) while the Young Guard comprised young men conscripted in the later years of Napoleon's reign. Exactly which of these groupings an individual guard regiment is considered part of is a matter of some debate and in all cases many different regiments could be taken as Old, Middle or Young Guard.

French Imperial Guard Grenadiers 1806-15

The 'Old Grumblers' of the Emperor's Imperial Guard (victors of 20 years of campaigning under General, Consul and Emperor) were destined to end their days in defeat at Waterloo. The Old Guard reserved their bearskins for the day of battle, and the victory parades, on the march they wore a bicorne decorated with aurore (orange) bands and the guard blue greatcoat. Towards the end of Napoleon's reign, with supplies overstretched, this marching order probably served for battle as well. Certainly by the time of the 1815 campaign, members of the so-called middle guard regiments (3rd and 4th Grenadiers and 3rd and 4th Chasseurs a pied) would have done well to have arrived at Waterloo so well equipped. The officer would be in common campaign dress throughout the period, officers rarely campaigning in the bearskin.

Unit of French Imperial Guard Grenadiers.

Imperial Guard Grenadiers Campaign Dress with Bearskin

The Old Guard Grenadiers campaigned in a variety of mixes of clothing during the period of the 'First Empire' (as Napoleon's time is known to the French). Known uniforms include:

Year	Uniform
1806-7	Full dress bearskin, surtout, breeches, black gaiters.
1809	Bearskin without plume or cords, surtout, and white trousers.
1813-14	Bearskin without cords or plume, full dress coat and blue trousers (worn over breeches and white gaiters, only the bottoms of the latter would be showing).
1815	Full dress bearskin, full dress coat, and white trousers (this was the dress described on their return from Elba, not necessarily what was worn on the Hundred Days campaign).

Old Guard Campaign Dress with Bicorne.

Old Guard in Campaign Dress.

Imperial Guard Grenadiers in Full Dress with Bearskin

'La Garde' in full dress, gaiters white or black, plumes, the lot – magnificent! These miniatures also suit the Italian Royal Guard and the Spanish Guard of Napoleon's brother, Joseph.

Imperial Guard Grenadiers in Full Dress with Bearskin.

Dutch Grenadiers of the French Imperial Guard 1810-1812

Incorporated into the Guard when Holland became part of municipal France, 'La Garde Hollandais' wore white uniforms faced red. Originally they had sported lace 'brandenbourgs' on their lapels, but this element was quickly dispensed with. Their greatcoat were a brighter medium blue compared to the French navy of the Old Guard Grenadiers and Chasseurs and for campaign they usually wore white trousers. These figures may be used for the Westphalian Guard Grenadiers as well. Dutch Grenadiers wore an identical uniform to the French, apart from the lack of a front plate on the bearskin; so Dutch Grenadiers in bicorne can be used as French and vice versa.

Dutch Grenadiers of the French Imperial.

The Young Guard

The Young Guard was the name given to a number of regiments raised later in Napoleon's reign, including regiments of tirailleurs (or sharpshooters), voltigeurs and flanqueur-grenadiers. Bardin made changes to the dress of light infantry as well as the line and, as these changes occurred at the same time as the formations of Young Guard infantry evolved, it is not surprising that there were many similarities in appearance between the two groups.

Unit of Young Guard Line Infantry.

By 1814 there were 16 regiments of tirailleurs and the same number of voltigeurs. This made the Guard the size of a large army corps but certainly weakened the overall quality of its manpower. Flanquer grenadiers and flanquer chasseurs were intended as light infantry support for the Old and Middle Guard regiments; they each consisted of two battalions of six companies. The tirailleurs and voltigeurs were dressed in the 'habit veste' coloured in a similar way to those of the light infantry. In addition they both had red pointed cuffs, piped white. The tirailleurs had red collars and the voltigeurs yellow ones, both being piped blue. Shakos were plain with brass 'eagle' plates and pompoms of red for tirailleurs and green for voltigeurs.

The 'flanquer' regiments both dressed in green with yellow piped uniforms of light infantry style. The chasseur's shako was plain but the grenadier's bore white Vs on either side in the style of the fusilier-grenadiers. Pompoms were yellow-over-red for the grenadiers and yellow-over-green for the chasseurs. Officers were normally drawn from the Old and Middle Guard regiments and often retained their blue uniforms.

Young Guard Flanquer Grenadiers & Flanquer Chasseurs.

French Cavalry

French Cuirassiers 1807-15

The strike force of Napoleon's cavalry, cuirassiers were big men on big horses, under some of the great cavalry generals of the First Empire – D'Hautpoul and Nansauty, who led the two elite divisions through the glory years of Empire, Espagne who took over after d'Hautpoul's death and Montbrun who had the style of Lasalle but none of his character. Indeed Lasalle himself died at Wagram, not with his beloved Hussars but leading the men who charged with d'Hautpoul.

The early cuirassiers were armed with a sword and pistols; the carbine was not issued until 1812. Until 1810 the Cuirassiers used only two different colours, red and yellow, to differentiate the regiments. This was possible using different types of pocket, horizontal or vertical, in combination with the facing colours appearing on various mixes of collar, cuffs and cuff-flaps. From 1810-1812 a new garment to be worn beneath the cuirass was issued, this 'habit-surtout' had no pockets thus rendering one of the methods of identification redundant. The way around this was to introduce two new facing colours aurore (orange) and pink. Finally, the 'habit-veste' arrived in 1812 but whilst the arrangement of the colours was altered the colours themselves remained the same. However two further regiments were created between 1808 and 1810 (13th & 14th) and these were given the facing colour 'lie de vin' or wine dregs (a sort of dark purply pink).

After 1812 Trumpeters, who normally had worn reversed

French Cuirassier Officer.

colours, were supplied with the Imperial livery. There appears to have been some interesting variations on the standard saddlery for officers – leopard-skin shabraques (saddle cloths) and plain pointed ones, like those used by hussar officers.

French Hussars 1807-15

Hussars were the light cavalry of the French (and many other European) armies. The most gloriously colourful element of La Grande Armée with their braided dolman and flying pelisse, the Hussars were often accused of 'faisant faquins', or showing off, when on campaign!

Many Hussar regiments had long, distinguished histories and noteworthy origins. The 1st were the old Bercheny regiment of the 'Ancien Regime', the 11th were in fact Dutch, and used cloth shabraques rather than sheepskin, while their elite company wore white colpacks, the 13th were raised twice, once from Italian recruits in 1813 and again from the Westphalian regiment of Jerome-Napoleon.

Several Hussar regiments played crucial roles in the wars of Napoleon's time. Sergeant Guindey of the 10th who killed Prinz Ferdinand of Prussia in single combat at Saalfeld in 1806; the 3rd fought alongside Auguste Colbert at Jena. Lasalle, known as 'the Hussar General' commanded perhaps the finest cavalry brigade of the age – 'La Brigade Infernale', comprising the 5th and 7th Hussar regiments.

Most of the states in the Confederation of the Rhine had, at one time or another, Hussars in the image of the French. Certainly this was the case in Westphalia, Bavaria and Saxony. Others included the Dutch-Belgians and, once again, the Italians and Neapolitans.

Regimental colours were as follows.

Regiment	Uniform
1st	Light blue and white.
2nd	Brown and pale blue.
3rd	Blue-grey and deep red.
4th	Blue and red.
5th	Blue dolman and white pelisses.
6th	Red and blue (the 4th's colours reversed).
7th	Green, red and gold.
8th	As the 7th, but with white braid.
9th	Pale blue and red.
11th	Cloth shabraques; white colpacks for their elite company.
12th	As the 9th, but with white rather than yellow braid.
13th	Raised twice. The first incarnation were dressed like the 2nd, the second in red and blue with yellow lace and red shakos.
14th	As the 8th.

Trumpeters generally, but not always, wore some form of reversed colours.

Hussars Elite Company.

French Dragoons

Dragoons were the workhorse cavalry of the French army, and with thirty regiments the most numerous apart from the Chasseurs. Nearly all the great cavalry leaders at one time commanded Dragoons; like Grouchy and Latour-Marbourg with great success, others such as Lasalle were to find their destiny elsewhere. The men wore medium green coats, which set off the facing colours of scarlet, crimson, pink, yellow and orange, differentiated by the usual: pockets, collars, cuffs and cuff-flaps. Like most French cavalry regiments the Dragoon had one company designated as elite, these being distinguished by the wearing of the bearskin. Italian Dragoons also wore the bearskin.

French Line Lancers

In 1811 six regiments of Dragoons (1st, 3rd, 8th, 9th, 10th and 29th) were converted into Chevau-leger Lanciers de la Ligne; previously the only lancers in the French army had been the two regiments of the Guard. The new regiments were issued short square fronted coats, in the same green as their old 'habits', hussar/chasseur-style green breeches with Hungarian boots. Their headgear was not unlike their old Dragoon helmet but bore an upright comb rather than the falling horse hair of their previous incarnation. For campaign wear the breeches were covered with overalls of green or grey reinforced with leather and with a strip of facing colour material down the outside opening which was secured with flat brass buttons. In the first year or so of their existence trumpeters were dressed at the colonel's instruction, but after 1812 this was replaced by the obligatory wearing of the new Imperial livery.

In the elite company the shoulder straps were replaced by red epaulettes, a red plume was worn and in certain regiments the helmet crest was coloured red.

CARABINIERS

Carabiniers were so named because they carried carbines. Initially the Carabiniers were uniformed as a type of Grenadier a Cheval (mounted grenadiers), forming part of Nansouty's division they fought in all the great campaigns. Trumpeters wore reversed colours of red faced blue.

In 1811 a new white uniform (it would have been red but for a shortage of dye) with brass helmet and cuirass was introduced; it was worn during the battles of the final years of Napoleon's Empire. The Trumpeters began wearing blue coat and helmet combs but were issued Imperial livery after 1812.

IMPERIAL GUARD CHASSEURS A CHEVAL

The Chasseurs a Cheval of the Imperial Guard were Napoleon's favourite cavalry regiment and acted as his personal body guard. Early on, the Imperial Guard cavalry would have appeared in full dress; from 1812 campaign dress became more common.

During 1813-14 three squadrons of Young Guard Chasseurs were formed. They were dressed similarly to the senior

squadrons apart from their headgear (a red cylindrical shako), their breeches (green with aurore decorations), and the colours of their shabraque (red with green decoration).

IMPERIAL GUARD POLISH & DUTCH LANCERS

The Poles were faithful allies of the French during the wars of Napoleon and in 1809 Napoleon formed a regiment of light horse to be part of the Imperial Guard. They were not

at this time equipped with the lance. They wore contrasting blue with pinky-crimson facings with white/silver lace. The Dutch joined later as a result of the incorporation of the Netherlands into France and wore red with blue facings and yellow/gold lace.

In campaign dress the facings are buttoned over and waterproof covers placed on the czapka. The Dutch also appear in blue overalls in contrast to the red they wear for

full dress. Trumpeters wear white-faced crimson in the Polish lancers and pale blue-faced red in the Dutch. No Guard regiment appeared to wear the Imperial livery at any time.

CHASSEURS

The chasseurs were the indigenous French light cavalry. They wore dark green with facings of scarlet, yellow, pink, crimson, orange, sky-blue, aurore, capucine, madder

red, amaranth, and chamois. Interesting variations of the dress include the 1st who wore a crested helmet during the One Hundred Days (they had been issued them during the Bourbon restoration as a Royal regiment), trumpeters of the 7th, who wore a pink Kinski coat and pink shako in 1808-9, as well as the 1st, who wore a cylindrical shako in 1814.

Elite companies were marked out by the addition of a colpack. Trumpeters generally wore reversed colours.

Chasseurs Elite Company

Napoleon & His Generals

Napoleon Bonaparte in his well worn garb of grey coat and undress uniform of the Guard Chasseurs a Cheval plus grey horse.

Marshal Murat as the King of Naples. Murat's clothing got more ridiculous as time passed and this outfit is probably one of the last tasteful ones before he entered the Russian campaign of 1812.

Marshal Ney, the bravest of the brave, whose behaviour during the campaign of 1812 was as serious and heroic as Murat's was comic and ultimately cowardly. After the Meisonnier painting '1814'.

Marshal Davout, the 'Iron Marshal', victor of Auerstadt in 1806 and perhaps the best of Napoleon's Marshals. Equipped with a pair of glasses for his poor eyesight.

Marshal Soult, Called by Napoleon the best strategist in Europe. Wellington's toughest opponent in Spain.

Marshal Bessieres Commander of Napoleon's Guard Cavalry, he 'lived like Bayard and died like Turenne' ripped apart by a cannon round at Rippach in 1813.

Marshal Massena. A hero of the Revolutionary wars, second only to the Emperor. Another candidate for the pick of the Marshals. Fell out with his Generals during the Peninsular War when he brought along his mistress dressed as a Hussar!

Napoleon seated, arms folded, foot on drum: the grumpy miserable genius of the later campaigns.

Roustan with the Emperor's telescope, the fat Mamaluke aide who was later to betray him.

Marshal Berthier with telescope. The best Chief of Staff in history who refused to serve in 1815 but fell from a window whilst watching the allies marching on France. Did he fall, did he jump or was he pushed?

Marshal Victor with telescope, wearing cape. Victor was one of the turncoats of 1814 and voted for the death penalty for Ney in 1815.

Marshal Grouchy in greatcoat with map. An excellent commander of cavalry, but out of his depth during the Waterloo campaign, Grouchy will always be remembered for failing to arrive on the battlefield of Waterloo. After the battle he fought an amazing withdrawing action to extricate his wing of the army from a perilous position.

Marshal Ney, hat in hand, pointing at map.

Marshal Mortier looking through telescope Poor old Mortier, not the brightest of Generals, who survived unscathed the wars of Napoleon only to be blown-up twenty years later by a bomb thrown at King Louis-Philippe!

General Savary, ADC to Napoleon, with messages from Napoleons spy master and 'spin doctor'.

Marquis of Vence, Imperial Orderly Officer. ADC, Hussar Uniform, pointing. Assistant to staff with document folder. Chasseur a Cheval of the Guard, dismounted on guard duty.

Napoleon *Roustan* *Berthier*

Victor *Grouchy* *Ney* *Staff*

Mortier *Savary* *Orderly* *ADC*

French Armies of the Peninsular War

The following army lists represent French armies from various stages of the war in Spain and Portugal. Their enemies during this time were, of course, the British, whose forces are described on page 149. During the same period of time, French armies were also fighting wars across Europe; army lists for these forces can be found elsewhere in this book.

French Army Special Rules

The following special rules apply to all French Peninsular armies:

National Contingents

Most French Peninsular War armies contained a number of soldiers of other nationalities (German, Italian and Swiss troops being the most common) as well as native French troops themselves, these nations can be deployed as either mixed or independent divisions.

Swiss Troops

The Swiss were known as particularly able soldiers. Swiss units are always classed as elite but an army may contain no more than one Swiss unit for every division in the army.

German Troops

German troops sometimes originated from Nassau, Berg, Westphalia and even Prussia and will normally be of line or militia status as they were often not keen on fighting for their French masters.

Italian & Polish Troops

Italian and Polish troops (Vistula units) are always classed as elite or line status.

Changing Formation

French armies were exceptionally well-drilled. All infantry units, except militia, can make a free formation change before moving in the Movement phase.

Mixed Columns

Three infantry units from the same division can form a mixed column, consisting of a unit of infantry deployed in line formation flanked on either side by a unit in column of attack.

French in Mixed Column.

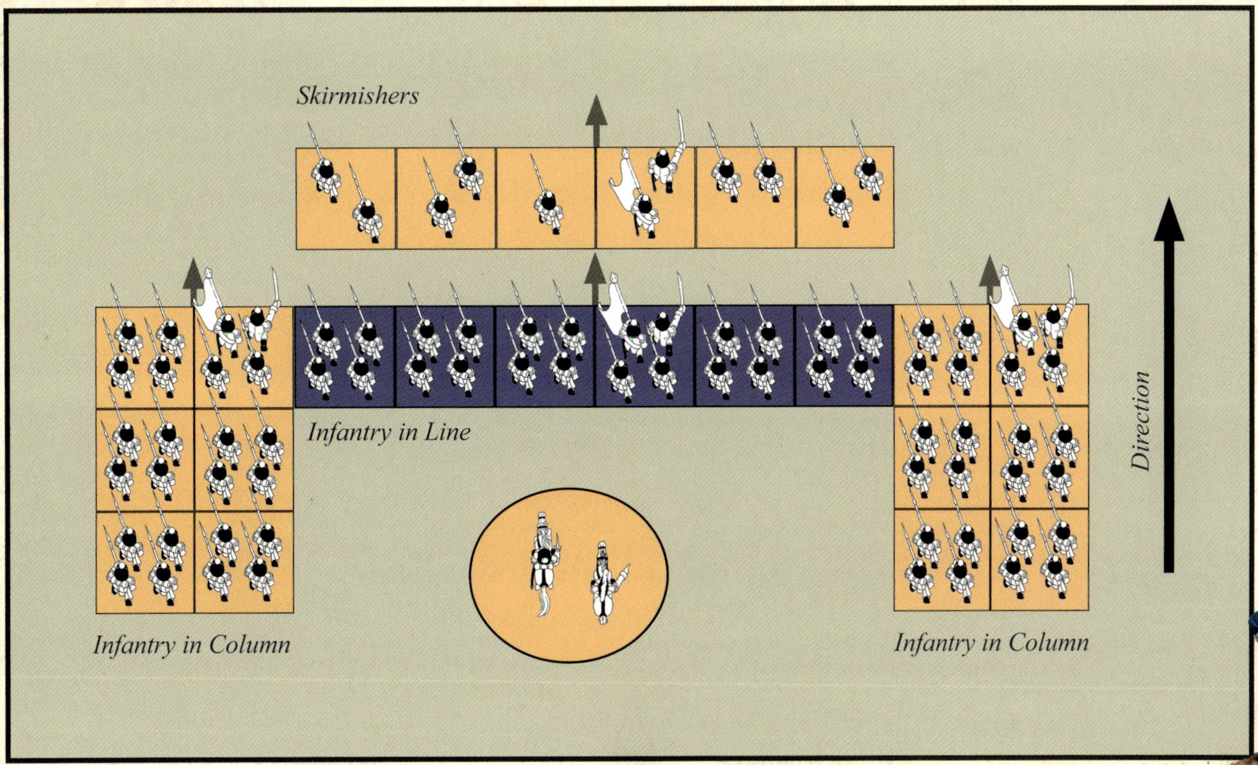

French Young Guard Officer.

French Chasseurs a Cheval.

Provided that none of the units in the mixed column have failed any morale tests, the three units are treated as one. The Divisional General has discretion whether to give the units commands individually or to give one command card to the whole formation. If the division in question contains any other units they are not treated as part of the mixed column.

Any artillery ball or howitzer shell fire is measured to the centre of the middle unit. When fired at, the French player may choose which of the formations in the mixed column will be used for the purposes of determining hits.

Divisional Square

Infantry units in a division containing four or more infantry units can form a divisional square. A unit in line formation is placed at the front and rear with columns of march on each side. All bases must be in direct contact. To form a divisional square, a change of formation is required, carried out in the normal manner. All units forming the divisional square will need to change formation, and all units must begin in contact with at least one other unit forming the divisional square in order to do so.

The Divisional Square formation can only make standard moves and any terrain obstacles will break the formation.

The Divisional Square formation counts as square if attacked by cavalry with one exception; on stand and fire orders, all sides being charged roll 6D6.

Artillery firing at the divisional square formation receives a plus 1 to hit when firing ball or canister.

Divisional Square.

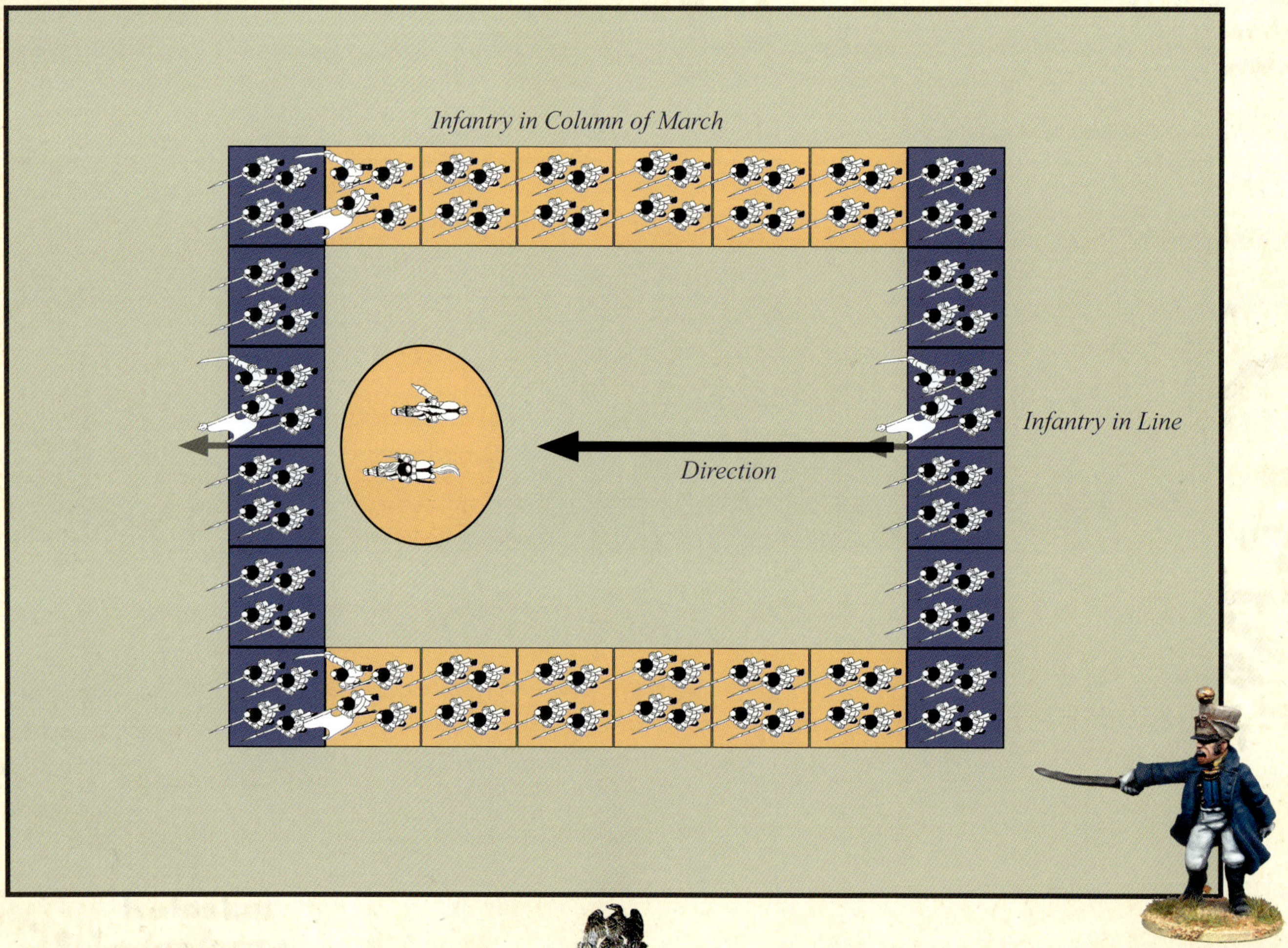

French Light Infantry Officer.

THE OLD GUARD

The Old Guard were the absolute epitome of a resolute veteran unit; Napoleon saved them as his reserve and their name alone struck terror into their opponents. In typical fashion they were the last to leave the disaster of Waterloo.

There is still conjecture as to which regiments actually formed the Old Guard but for these rules, the matter has been simplified. The Old Guard contains the following troops: the 1st and 2nd Grenadiers a Pied, 1st and 2nd Chasseurs a Pied, Chasseurs a Cheval, Grenadiers a Cheval, Gendarmerie d'Elite, Empress Dragoons, Polish and Dutch Lancers, Mamelukes, Engineers, Seamen and all Guard Artillery. All other middle Guard formations count as Guard status; Young Guard units count as elite status (all of which is detailed in the army lists). The following special rules apply specifically to the Old Guard themselves.

Baron Larry's Ambulance

Any Old Guard infantry units can have a Baron Larry's ambulance attached at a cost of +5 points. Any unit with the ambulance attached rolls 1D6 each turn and on a roll of 6 the number of casualties received that turn is reduced by one.

Fearsome Reputation

Enemy units, with the exception of Guard units, must pass a morale test to charge an Old Guard unit. The Old Guard unit does not need to stand and fire. If the charging unit fails its morale test it the charge fails.

Old Guard Generals

Divisional Generals and Brigade commanders in Old Guard units never have a Command Rating that is never lower than the score in the chart. Any rolls lower than on the chart are considered to be the same as the chart.

GRENADIERS

By the time of the Napoleonic Wars, French armies seldom fielded brigades or units of grenadiers. Instead, smaller grenadier companies fought as part of line infantry divisions.

French Hussar.

Occasionally, however, these grenadier companies would be detached from the other line infantry to form units of their own, particularly if the army was lacking in elite infantry. Where Grenadiers are included in the army list, they are not purchased as other units are, but instead are drawn from amongst the line infantry.

If grenadiers are available in the army list, up to one unit of grenadiers is allowed for every four units of line infantry in the army. Each of these four line infantry units has one of its bases removed (thus reducing its unit size to 5) and these bases are placed together to form a unit of grenaders (consisting of 4 bases). The grenadier unit is elite, but the depleted line infantry units have their firing in line reduced to 5D6, and a reduced morale test of 2D6. The grenadier units themselves also fire with 5D6 when firing in line. If the army contains three or more grenadier units, they will form a division in their own right, with a divisional general to lead them (who is included at no extra points cost). Otherwise, the grenadier unit forms part of the division from which its members are drawn. The inclusion of grenadier units has no effect on your army's infantry PSV.

CUIRASSIERS

Retreating cuirassiers ignore the first extra hit suffered in each round of combat.

SKIRMISHERS

All French infantry units contained a high proportion of skirmishers. To represent this each infantry unit (except militia) can fire 2D6 worth of skirmisher fire as long as the unit is not in square or column of march. See Skirmisher Fire (page 53) for range and effect.

TACTICAL SUPERIORITY

French armies constantly confused their opponents with their unexpected tactical movement, speed and imagination, the Prussians in particular never seemed to be able to cope with this!

After both armies have deployed the French commander can immediately move some, or all, of his infantry divisions.

For each infantry division roll 1D6 and consult the following chart:

D6 Score	Division Movement
1	The division receives no free moves.
2-3	The division receives 1 free move.
4-5	The division receives 2 free moves.
6	The division receives 3 free moves.

When fighting Prussian armies, roll 2D6 and choose the best score. These free moves may be used to make standard moves or march moves (maximum two march moves). They may not be used to force march. Movement is not compulsory but any movements not used are lost.

Early Peninsular War (1807-08) French

Command Ratings

French command ratings in the early Peninsular War should be reduced by a score of one for divisional generals and by a score of two for brigade commanders. The commander in chief's command rating is unaffected.

Divisions, Brigades & Batteries

Your army may include:

Type	Availability	Consisting of:
Infantry Division	Your army may include any number of infantry divisions.	3-5 infantry units, plus a divisional general. Up to two units may be light infantry; up to three units may be elite (either light or line).
Foot Artillery Battery	Your army may include one foot artillery battery for every six infantry units in the army. If your army has five infantry units or less you may choose a half battery at half the points cost.	3 foot artillery guns & 1 howitzer. Half battery consists of 2 foot artillery guns.
Horse Artillery Battery	Your army may include one horse artillery battery for every four cavalry units in the army.	2 horse artillery guns & 1 howitzer.
Cavalry Brigade	Your army may include one cavalry brigade for every infantry division in the army.	1 or 2 units of cavalry, plus a brigade commander; both units must be of the same type.
Cavalry Division	Two cavalry brigades may be combined into a cavalry division. You must meet the requirements for fielding both brigades.	As brigades – may be one heavy cavalry brigade and one light cavalry brigade, or may be all of one type. One brigade commander is promoted to divisional general.
Grenadier Units	Your army may include one unit of grenadiers for every four line infantry units in the army.	One line infantry base is removed from each unit; these bases form a unit of grenadiers. See special rules.

Units

Choose the units that make up your divisions from the following:

Troop Type	Points Cost	Unit Size	Morale/Class
Line Infantry	20 points per unit	6	Line Infantry
Light Infantry	25 points per unit	6	Line Infantry
Combined Grenadiers	Free	4	Elite Infantry
Elite Line Infantry	25 points per unit	6	Elite Infantry
Elite Light Infantry	30 points per unit	6	Elite Infantry
Light Cavalry	35 points per unit	6	Light Cavalry
Dragoons	40 points per unit	8	Heavy Cavalry
Foot Artillery Battery	60 points/30 points	3 foot artillery & 1 howitzer	Foot Artillery
Horse Artillery Battery	75 points	2 horse artillery & 1 howitzer	Horse Artillery

The keys of a fortress are always well worth the retirement of the garrison when it is resolved to yield only on those conditions. On this principle it is always wiser to grant an honourable capitulation to a garrison which has made a vigorous resistance than to risk an assault.

— Napoleon Bonaparte

French Light Infantry.

Peninsular War (1808-09) French

Divisions, Brigades & Batteries

Your army may include:

Type	Availability	Consisting of:
Infantry Division	Your army may include any number of infantry divisions.	3-5 infantry units, plus a divisional general. Up to two units may be light infantry; up to four units may be elite (either light or line).
Foot Artillery Battery	Your army may include one foot artillery battery for every six infantry units in the army. If your army has five infantry units or less you may choose a half battery at half the points cost.	3 foot artillery guns & 1 howitzer. Half battery consists of 2 foot artillery guns.
Horse Artillery Battery	Your army may include one horse artillery battery for every two cavalry brigades in the army.	2 horse artillery guns & 1 howitzer.
Cavalry Brigade	Your army may include one cavalry brigade for every infantry division in the army, up to a maximum of three.	1 or 2 units of cavalry, plus a brigade commander; both units must be of the same type.
Guard Infantry Division	Your army may include one guard infantry division if it includes at least two infantry divisions.	2-4 Old Guard infantry units, plus an Old Guard divisional general.
Guard Light Cavalry Brigade	Your army may include one guard light cavalry brigade if it includes a guard infantry division or four or more infantry divisions.	1-3 guard cavalry units, plus an Old Guard divisional general (not a brigade commander).
Guard Foot Artillery	You may upgrade one battery of foot artillery to guard artillery if you army includes a guard infantry division.	This is an upgrade and does not bring your army any additional units: one foot artillery battery gains guard status and all foot artillery in the battery becomes heavy artillery.
Guard Horse Artillery	You may upgrade one battery of horse artillery to guard artillery if you army includes a guard cavalry division.	This is an upgrade and does not bring your army any additional units: one horse artillery battery gains guard status.

French confront the British in Spain.

UNITS

Choose the units that make up your divisions from the following:

Troop Type	Points Cost	Unit Size	Morale/Class
Line Infantry	20 points per unit	6	Line Infantry
Light Infantry	25 points per unit	6	Line Infantry
Elite Line Infantry	25 points per unit	6	Elite Infantry
Elite Light Infantry	30 points per unit	6	Elite Infantry
Old Guard Infantry	40 points per unit	6	Old Guard Infantry
Light Cavalry	35 points per unit	6	Light Cavalry
Dragoons	40 points per unit	8	Heavy Cavalry
Guard Light Cavalry	50 points per unit	6	Guard Light Cavalry
Foot Artillery Battery	60 points/30 points	3 foot artillery & 1 howitzer	Foot Artillery
Horse Artillery Battery	75 points	2 horse artillery & 1 howitzer	Horse Artillery
Guard Foot Artillery	+30 points	-	Guard Heavy Artillery
Guard Horse Artillery	+15 points	-	Guard Horse Artillery

Battle of Talavera July 1809.

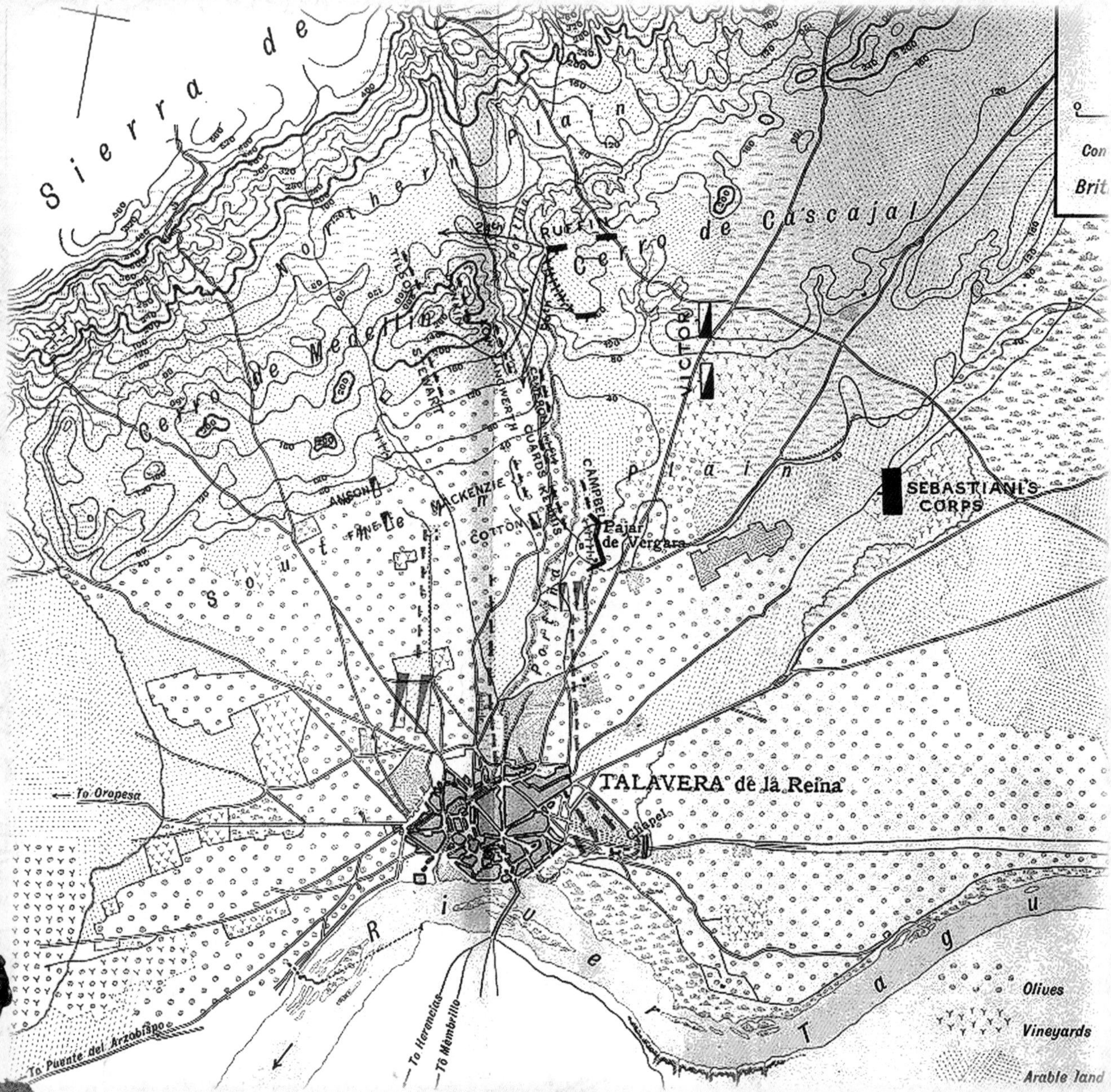

French Grenadier.

Peninsular War (1809-10) French

Command Ratings

French command ratings in the early Peninsular War should be reduced by a score of one for divisional generals and by a score of two for brigade commanders. The commander in chief's command rating is unaffected.

Divisions, Brigades & Batteries

Your army may include:

Type	Availability	Consisting of:
Infantry Division	Your army may include any number of infantry divisions.	3-5 infantry units, plus a divisional general. Up to two units may be light infantry; up to two units may be elite (either light or line). Up to half the infantry in your army may be militia.
Foot Artillery Battery	Your army may include one foot artillery battery for every five infantry units in the army. If your army has four infantry units or less you may choose a half battery at half the points cost.	3 foot artillery guns & 1 howitzer. Half battery consists of 2 foot artillery guns.
Horse Artillery Battery	Your army may include one horse artillery battery for every two cavalry brigades in the army.	2 horse artillery guns & 1 howitzer.
Cavalry Brigade	Your army may include one cavalry brigade for every infantry division in the army, up to a maximum of four.	1 or 2 units of cavalry, plus a brigade commander; both units must be of the same type.
Cavalry Division	Your army may include one cavalry division for every three infantry divisions in the army. The cavalry division counts as two cavalry brigade choices.	Chosen as two separate brigades – may be one heavy cavalry brigade and one light cavalry brigade, or may be all of one type. A cavalry division must include at least three units. One brigade commander is promoted to divisional general.
Grenadier Units	Your army may include one unit of grenadiers for every four line infantry units in the army.	One line infantry base is removed from each unit; these bases form a unit of grenadiers. See special rules.

Units

Choose the units that make up your divisions from the following:

Troop Type	Points Cost	Unit Size	Morale/Class
Militia	15 points per unit	6	Militia Infantry
Line Infantry	20 points per unit	6	Line Infantry
Light Infantry	25 points per unit	6	Line Infantry
Combined Grenadiers	Free	4	Elite Infantry
Elite Line Infantry	25 points per unit	6	Elite Infantry
Elite Light Infantry	30 points per unit	6	Elite Infantry
Light Cavalry	35 points per unit	6	Light Cavalry
Lancers	40 points per unit	6	Light Cavalry
Dragoons	40 points per unit	8	Heavy Cavalry
Foot Artillery Battery	60 points/30 points	3 foot artillery & 1 howitzer	Foot Artillery
Horse Artillery Battery	75 points	2 horse artillery & 1 howitzer	Horse Artillery

PENINSULAR WAR (1811-12) FRENCH

COMMAND RATINGS

French command ratings in the early Peninsular War should be reduced by a score of one for divisional generals and by a score of two for brigade commanders. The commander in chief's command rating is unaffected.

DIVISIONS, BRIGADES & BATTERIES

Your army may include:

Type	Availability	Consisting of:
Infantry Division	Your army may include any number of infantry divisions.	3-5 infantry units, plus a divisional general. Up to two units may be light infantry; one unit may be elite (either light or line). Up to half the infantry units in your army may be militia.
Young Guard Infantry Division	Your army may include one Young Guard infantry division if it includes at least two infantry divisions.	2-4 elite infantry units, plus a divisional general; up to two units may be elite light infantry.
Foot Artillery Battery	Your army may include one foot artillery battery for every five infantry units in the army. If your army has four infantry units or less you may choose a half battery at half the points cost.	3 foot artillery guns & 1 howitzer. Half battery consists of 2 foot artillery guns.
Horse Artillery Battery	Your army may include one horse artillery battery for every two cavalry brigades in the army.	2 horse artillery guns & 1 howitzer.
Cavalry Brigade	Your army may include one cavalry brigade for every infantry division in the army, up to a maximum of three.	1 or 2 units of cavalry, plus a brigade commander; both units must be of the same type. For armies of 1812, up to half the light cavalry in your army may be lancers; lancers may be fielded in brigades of their own or mixed with other light cavalry.
Grenadier Units	Your army may include one unit of grenadiers for every four line infantry units in the army.	One line infantry base is removed from each unit; these bases form a unit of grenadiers. See special rules.

UNITS

Choose the units that make up your divisions from the following:

Troop Type	Points Cost	Unit Size	Morale/Class
Militia	15 points per unit	6	Militia Infantry
Line Infantry	20 points per unit	6	Line Infantry
Light Infantry	25 points per unit	6	Line Infantry
Elite Line Infantry	25 points per unit	6	Elite Infantry
Elite Light Infantry	30 points per unit	6	Elite Infantry
Light Cavalry	35 points per unit	6	Light Cavalry
Lancers	40 points per unit	6	Light Cavalry
Dragoons	40 points per unit	8	Heavy Cavalry
Foot Artillery Battery	60 points/30 points	3 foot artillery & 1 howitzer	Foot Artillery
Horse Artillery Battery	75 points	2 horse artillery & 1 howitzer	Horse Artillery

Peninsular War (1813-14) and the Allied Invasion of South-western France French

Command Ratings

French command ratings in the early Peninsular War should be reduced by a score of one for divisional generals and by a score of two for brigade commanders. The commander in chief's command rating is unaffected.

Divisions, Brigades & Batteries

Your army may include:

Type	Availability	Consisting of:
Infantry Division	Your army may include any number of infantry divisions.	3-5 infantry units, plus a divisional general. Up to two units may be light infantry Any number of units may be militia.
Young Guard Infantry Division	Your army may include one Young Guard infantry division if it includes at least two infantry divisions.	2-4 elite infantry units, plus a divisional general; up to two units may be elite light infantry.
Foot Artillery Battery	Your army may include one foot artillery battery for every five infantry units in the army. If your army has four infantry units or less you may choose a half battery at half the points cost.	3 foot artillery guns & 1 howitzer. Half battery consists of 2 foot artillery guns.
Horse Artillery Battery	Your army may include one horse artillery battery if the army includes two cavalry brigades.	2 horse artillery guns & 1 howitzer.
Cavalry Brigade	Your army may include one cavalry brigade for every infantry division in the army, up to a maximum of two.	1 or 2 units of cavalry, plus a brigade commander; both units must be of the same type. One unit in each light cavalry brigade may be lancers..

Units

Choose the units that make up your divisions from the following:

Troop Type	Points Cost	Unit Size	Morale/Class
Militia	15 points per unit	6	Militia Infantry
Line Infantry	20 points per unit	6	Line Infantry
Light Infantry	25 points per unit	6	Line Infantry
Elite Line Infantry	25 points per unit	6	Elite Infantry
Elite Light Infantry	30 points per unit	6	Elite Infantry
Light Cavalry	35 points per unit	6	Light Cavalry
Lancers	40 points per unit	6	Light Cavalry
Dragoons	40 points per unit	8	Heavy Cavalry
Foot Artillery Battery	60 points/30 points	3 foot artillery & 1 howitzer	Foot Artillery
Horse Artillery Battery	75 points	2 horse artillery & 1 howitzer	Horse Artillery

The first qualification of a soldier is fortitude under fatigue and privation. Courage is only the second; hardship, poverty, and want are the best school for a soldier.

— Napoleon Bonaparte

French Infantryman.

Charges of cavalry are equally useful at the beginning, the middle and the end of a battle. They should be made always, if possible, on the flanks of the infantry, especially when the latter is engaged in front.

— Napoleon Bonaparte

French Camel Regiment (Knötel).

THE BRITISH ARMY

UNIFORMS AND COLOURS

BRITISH LINE INFANTRY 1809-15

The backbone of the British armies in the Peninsular War was the line infantry, whose resurrection from a series of embarrassing defeats in the preceding decades began just days after their arrival in Portugal as they inflicted two defeats on the 'invincible' French.

Unit of early British Line Infantry, centre company.

Wellington himself cared little for the turnout of his troops as long as they were fed and well supplied. However troops started out dressed in the early nineteenth century's best regimentals, the famous red coats with a range of facings: blue, buff, gosling green, deep yellow, light yellow, green, pale yellow, willow green, bright yellow, yellow, white, deep green, black, yellowish buff, yellowish green, red and deep buff.

The coat had loops down the front in the regimental lace; sergeants wore white, either singly or in pairs with the ends square or bastion shaped. The headgear was the ubiquitous 'stovepipe' shako, with a white-over-red plume for the centre companies. The centre company men were also recognisable by the simple white tufts at the end of their shoulder straps, which were facing colour edged with the regimental lace. The white breeches and short black gaiters were often covered (or replaced) by trousers of white, although lengthy service in Portugal and Spain added various shades of brown and grey to the ensemble. Later in the war trousers of grey or blue-grey became official wear. Officers wore long coats and large bicornes in the centre companies.

Early British Line Infantry marching in column.

British Line Infantry Flank Companies 1809-14

The dress of flank companies was different from the centre in two major respects: firstly the plumes on the shako were coloured white for Grenadiers and green for the Light Company and secondly the shoulder tufts were replaced by wings, coloured red and decorated with the regimental lace. Officers of the Light Company wore short coats with often elaborate wings, together with the shako, rather than the bicorne and light infantry NCOs carried muskets instead of pikes.

British Line Infantry Centre Companies 1812-15

In 1812 a new shako was authorised for the British infantry. However problems with supply meant that it is unlikely that it saw much use during the Peninsular war. By the time of Waterloo it's probable that all battalions, with the exception of the 28th, who had the honour of retaining the old shako, had received it. It differed from the old in having a false front, a side plume, cords and a new style front plate. The cords were white for the Centre and Grenadier companies. Grey trousers were now official campaign wear and officers, of all companies, were now issued with the new shako.

Late British Line Infantry in line.

BRITISH INFANTRY IN FATIGUE CAPS 1809-1814

Like the cavalry watering cap, the infantry fatigue cap was used on the march or in situations where the shako was lost or destroyed.

British Infantryman in Fatigue Cap.

HIGHLANDERS 1806-1815

The dress of Highlands regiment became more standardised in the early nineteenth century and in 1806 the following regiments were wearing it: 42nd, 71st, 72nd, 73rd, 74th, 75th, 78th, 79th, 91st, 92nd and 93rd.

Units that saw active service in the kilt consisted of:

Regiment	Service
42nd	Peninsular War 1808-9 and 1812-14, Walcheren 1809, Waterloo 1815.
71st	Argentina 1806.
78th	Maida 1806, Holland and Belgium (though they did not serve at Waterloo).
79th	Copenhagen 1807,Peninsular War 1808-9 and 1810-14, Walcheren 1809, Waterloo 1815.
92nd	Copenhagen 1807, Peninsular War 1808-9 and 1810-14, Walcheren 1809.

In 1809-10 most of the regiments were de-kilted as it was thought that they had a better chance of recruiting non highlanders if they were garbed in non-highland dress; only the 42nd, 78th, 79th, 92nd and 93rd retained the kilt, although the 93rd appeared to have worn trews at New Orleans in 1815.

Centre Company in Kilt & 'Mounted' Bonnet

The uniform coat of highland regiments was slightly shorter than those worn by other infantry but otherwise the positioning of lace, shoulder tufts and so on was as described above for the rest of the line infantry. The company plumes, worn on the bonnet, were for the most part coloured like those of the 'English' line regiments; however the 42nd were the exception as their centre companies wore the famous 'red hackle'(, a red plume), whilst their grenadiers wore red-over-white and their Light company red-over-green.

Highlander Flank Company in Kilt & 'Mounted' Bonnet

Highland flank companies had the same distinctions as other line infantry flank companies (see notes on 42nd above). Although Highland regiments were no longer viewed as some kind of light infantry, (as they had been during the mid-eighteenth century), they still retained the habit of using their Light Company troops as genuine skirmishers. This was frowned upon in other line regiments, as it was thought best to leave this aspect of battle to the 'specialists'!

Highlanders on the march.

BRITISH 95TH RIFLEMEN 1808-1815

Famously equipped with the Baker rifle, the 'sweeps' – as the 95th were known on account of the dark clothing and black facings, making them look like chimney sweeps – played a vital part in the victory in the Peninsular War and went on to serve at Waterloo. Trained and developed – like the rest of Crauford's famous light division – they served in some way in almost every major engagement of the war. Unlike the 60th, the other rifle regiment, they were usually deployed as a battalion, rather than parcelled out in company strength to serve as skirmishers to individual divisions. Their articulate officers and other ranks put together more diaries and memoirs of the wars than the rest of the army put together and it is from their accounts that tales of their appearance derive.

95th Riflemen.

Riflemen in action.

British Hussars 1806-12

It was with some reluctance that 'Horse Guards' authorised the conversion of three, then four, regiments of Light Dragoons into Hussars in 1806-7. This traditionalism showed further in the fact that in two of these cases the title Hussars was merely attached in brackets after that of Light Dragoons.

All of the Hussar regiments were to serve in the Peninsular War and on their first appearances they were wearing the somewhat impractical tall fur cap. The four regiments, at this time were carrying the following facing and other details (all units had white lace):

Regiment	Facings & Details
7th	Brown caps with red bags, white facings.
10th	Brown caps with yellow bags, yellow facings.
15th	Brown caps with red bags, red facings.
18th	Black caps with blue bags.

By 1812 the details had changed as the 10th had been granted red facings. In the light cavalry up until 1812, trumpeters wore either some form of reversed colours or some other mix of hues intended to differentiate them from the other ranks.

Hussars in Tall Fur Caps.

British Hussars in Fur Cap & Slung Pelisse 1812-15

At Waterloo, the Hussars had received a new improved version of the fur cap. Shorter and more comfortable it was worn by the 7th, 15th and 18th while the 10th wore the shako. Facings had changed again, for while those of the 15th and 18th had remained the same throughout, the braid of the uniform of the 7th and 10th had changed to yellow with their facing becoming blue.

British Hussars in Shako & Pelisse 1812-15

Although the shako was used at one time or another by most Hussar regiments, only the 10th wore it, in scarlet, at Waterloo. The black shakos shown here would have been typical of a number of regiments in the later part of the Peninsular War.

British Light Dragoons

Light dragoons fulfilled the role of light cavalry. The British light dragoons entered the Peninsular War in a similar form of dress to the Hussars. A lace jacket not unlike the Hussar

dolman was worn with overalls and the Tarleton helmet. Lace was either white or yellow and facing colours were pale buff, pale yellow, yellow, orange and scarlet.

In 1812 a new uniform was introduced which replaced the Tarleton with a shako and the laced jacket with a French style habit with coloured facings on the front. Indeed the whole ensemble was so 'frenchified' that it was much detested in the army. Both these alterations and those made to the heavy cavalry brought with them the danger of British troops being mistaken for French, something deplored by many, including Wellington.

BRITISH DRAGOONS

The British heavy cavalry served through most of the war in Spain wearing a braided red coat and a large bicorne. The latter was rather impractical, so as they began to deteriorate they were replaced by the shako.

The shako (or watering cap) was used as a stop gap to make the dress of heavy cavalry more practical. It had been issued earlier for certain orders of dress and activities but it was used a great deal during the Spanish campaigns as the headgear of choice.

British Dragoon & Dragoon Guards in Helmets 1812-15

In 1812 a new uniform was introduced for the heavy cavalry. The bicorne was replaced with a new helmet of leather and brass (with a horsehair tail of similar style to that worn by French Dragoon) and a new coat with vertical bands of lace down the front. The idea of a new helmet had been mooted for some time and various types hat been experimented with, including one with a caterpillar crest striped black and red; this was rejected for use by the line regiments of heavy cavalry but, ironically, was adopted, albeit with a different colour scheme, by the Life Guards and the Royal Horse Guards!

Like the previous coat, there was little to distinguish between Dragoon Guards and Dragoons, particularly as the cuffs, which were one of the differences, were covered by gauntlets. However the other variation could be seen, with the Dragoons Guards having a patch of red cloth on the collar whereas on the Dragoons the lace of the coat front continued onto the collar.

2nd (Royal North British) Dragoons 'Scots Greys' 1812-1815

The two distinctive features of the Greys – their grenadier headdress and grey horses – were both retained after the changes to the uniform in 1812. The coat was faced blue, for a Royal regiment, and it is known that the 'Greys' wore black waterproofed covers over their bearskins at Waterloo (the regiment did not see service in the Peninsular War). Trumpeters in British heavy cavalry regiments wore the same uniform as the other troopers.

2nd (Royal North British) Dragoons 'Scots Greys'.

British Royal Artillery 1809-1814

The artillery followed the infantry in its style of dress; it fielded the stovepipe shako in Spain and Portugal and changed to the new 'Belgic' shako in time for Waterloo. The rest of the uniform consisted of a blue coat laced yellow and faced red and grey campaign trousers, plumes were white-over-red.

The Horse Artillery wore a version of the early Light Dragoon uniform, with white/silver braid and red facings and the Tarleton helmet.

Drivers for both Horse and Foot Artillery wore similar uniforms based on that worn by the Royal Horse Artillery.

Deployed British Horse Artillery.

British Heavy Dragoon.

British Officers

Duke of Wellington 'Old Nosey', plain saddle cloth, in his plain blue riding coat, as worn at Waterloo, on Copenhagen, his favourite horse.

Lieutenant-General Sir Thomas Picton in round hat and coat, plain saddle cloth A hero of the wars in Spain and Portugal, Picton was notorious for wearing any old mix of clothing, most of which was not military. His staff enjoyed the same reputation. He was killed leading his division at Waterloo.

Lieutenant-General Sir. Rowland Hill. ('Daddy' Hill) was one of Wellington's best and most trusted subordinates, he was well liked by his men whose welfare he was particularly concerned with.

Lieutenant-General the Earl of Uxbridge wearing pelisse, leopard-skin saddle cloth. An excellent General of Cavalry, Uxbridge commanded the cavalry in the Peninsular until Corunna. After committing the faux pas of running off with Wellington's sister-in-law, he was not employed again until the 1815 campaign. He lost a leg at Waterloo.

Captain Von Streerwitz, 2nd K. G. L. Hussars, ADC to Uxbridge. Adjutant-General Sir Edward Barnes standing, leaning on sword.

Major-General Sir Denis Pack standing reading map. Served bravely throughout the Peninsular War, having been wounded on eight occasions. Commanded a brigade at Waterloo and was wounded again.

Major-General Sir James Kempt standing, hat under arm, frock coat Commanded a brigade at Maida 1806 and later served in Portugal and Spain. Badly wounded at Badajoz and again at Nivelle. At Waterloo commanded the 5th Division after Picton's death.

Captain Von Streerwitz — *Lieutenant-General the Earl of Uxbridge* — *Duke of Wellington* — *Lieutenant-General Sir Rowland Hill* — *Lieutenant-General Sir Thomas Picton*

Major-General Sir Denis Pack — *Dragoon Guards Officer* — *ADC* — *Major-General Sir James Kempt* — *Major-General Sir Edward Barnes*

'Old Nosey' commands a British square at the height of the battle!

British and Allied Armies of the Peninsular War

The British army did not perform well during the Revolutionary Wars, with many of the lessons learned about the effectiveness of light troops in the American Revolution forgotten or considered unnecessary for warfare on the continent. The British army was also very small and Militia and Yeomanry formations were organised all over Britain in an attempt to fend off an expected French invasion.

Under Sir John Moore a few regiments were eventually trained to fight with light infantry tactics, the 60th and 95th Rifle regiments armed with the deadly Baker rifle being the most famous.

The Royal Navy's victory over the French and Spanish Fleet at Trafalgar finally erased the possibility of a French invasion, but Britain's naval superiority did not afford her an effective means of defeating Napoleon on land. In 1808, following a number of unsuccessful landings in other parts of Europe, a small expeditionary force was sent to Portugal under the command of General Wellesley. Wellesley quickly defeated the French at the battles of Rolica and Vimerio, but when more superior officers arrived on the scene they inexplicably allowed the French army to leave.

The following year, and after a heroic retreat, the British army under Sir John Moore was driven from Spain by French forces under Napoleon's direct command; Napoleon left Spain before the British army was completely destroyed – as events in Europe more urgently required his attention – leaving the completion of the task to Marshal Soult. Sir John Moore died in the final phase of the retreat.

The population of Spain instituted a bloody guerrilla war against the French invaders that disrupted French lines of communication and occupied the attention of tens of thousands of French troops. The British, under the reinstated Wellesley, took the opportunity the Spanish situation afforded to fortify Portugal, thereby giving the British a secure base from which to launch attacks on Spain. Finally, after six more years of fighting and a series of brilliant victories, the British, Portuguese and Spanish armies defeated Marshal Soult at the Battle of Toulouse and the Peninsular Campaign was over.

The following special rules apply to the armies of Britain and her allies:

British and Allies Special Rules

British Cavalry

British cavalry (excluding the King's German Legion and Brunswick Cavalry) were notorious for 'galloping at everything' and so their command rating is always reduced by two unless the unit is charging.

Portuguese Cavalry

Portuguese cavalry were of poor quality and are always considered to be militia cavalry.

Riflemen

Riflemen can be deployed in detachments (this is a departure from the standard rules for light infantry on page 81). Detachments are represented by six models based in pairs on 40mm by 40mm bases.

Riflemen have no parent unit but must stay within 8" of an allied infantry unit at all times. For firing purposes the nearest unit to the rifle detachment counts as its parent unit. The detachment calculates its own morale score using 2D6. Rifle detachments do not count when calculating your army's PSV but full rifle regiments do.

You can field one detachment of Riflemen per infantry division.

You can also replace one light infantry unit with a regiment of rifles. This is a unit of 6 bases of riflemen. The same rules apply as for light infantry except the whole unit is armed with rifles.

Highlanders

Highlanders were feared for their ferocious charges. Any kilted highland units charging into close combat can re-roll any scores of 1.

The King's German Legion Dragoons

King's German Legion Dragoons were of exceptional quality, carrying out many successful charges. If a King's German Legion unit charges more than a standard move they can re-roll any failed hits in close combat once per turn.

British & Allied Artillery

Allied artillery batteries contain six guns except for British half artillery batteries which should be represented by two gun models.

All cannon can fire shrapnel (an explosive shell similar to howitzer fire) and all the guns in a battery must fire the same type of shot, either ball, shrapnel or canister.

Shrapnel fire is worked out with the same firing rules as for ball, but the damage is calculated in the same way for the howitzer shell.

Column of Attack

Infantry cannot form column of attack formation.

Elites

Unless otherwise stated up to half of any division can be upgraded to elite status.

Scouting, Deployment & Terrain

Allied armies tended to fight more defensively and often possessed better intelligence on enemy dispositions and battlefield terrain. To represent this allied armies can deploy a division first before the enemy, irrespective of scouting points.

Deployment Cards

Allied armies can deploy units behind hills or ridges and, while these troops are behind the hill, the models representing them are not placed on the tabletop; instead a deployment card is placed face down to represent each unit. The card should be placed following the normal rules for deployment.

Deployment cards are placed face down upon the table; your opponent will be know that they are there but will not know which unit is present – we can assume that the enemy is only able to see dust clouds from troop movements. All Deployment cards have a frontage of 250mm and depth of 100mm regardless of the actual formation it represents.

Deployment cards can represent:

- Infantry in Line
- Infantry in Square
- Infantry in Column of March
- Light Infantry Deployed
- Cavalry in Line
- Cavalry in Column of March
- Cavalry in Column of Attack
- Artillery Limbered
- Artillery Unlimbered

Choose the appropriate card for the unit you wish to deploy, bearing in mind the formation, and place the card face down in the chosen spot on the table.

Revealing Deployment Cards

Each deployment card describes the unit type and its formation; as long as they remain behind the hill the unit cannot be shot at (except by speculative howitzer fire). As soon as the unit moves from behind to the front half of the hill they are considered to have been deployed and accordingly the unit's models are placed on the tabletop.

If any enemy troops, including skirmishers, move to the back half of the hill the hidden unit will deploy. In these circumstances the rules for engagement range do not apply, so it is possible for units on deployment to be very close.

As soon as the enemy unit has reached the back half of the hill it will halt instantly on the appearance of the allied troops who are now considered to be in engagement range. As soon as models are deployed they adopt the proper formation. Any hidden allied units formed in line are automatically given a Stand and Fire! order when they deploy (which you can change next turn).

It is very important that players agree on the exact position and extent of hills and ridges before the game starts in order to avoid any confusion (or argument) over which troops are counted as being hidden.

Volley Fire – Two-man Line

The firepower of the famous British two-man line was devastating. When British infantry units fire their first volley or fire at any time without having moved that turn (including Stand and Fire! orders) the line can roll 7D6 instead of the usual 6D6; Portuguese troops of 1812, 1813 and 1814 also receive this additional dice.

Allied Rifles

Portuguese Cacadores, King's German Legion light infantry and Brunswick Oels infantry units and detachments are armed with rifles.

Due to the relatively small size of Peninsular engagements compared to Western European battles (where these rules normally treat a group of around four units as a division), each group of units represents a brigade. This has no impact upon the game, however it does affect army selection.

At brigade level you cannot mix British and Portuguese troops in the same brigade, meaning wargames divisions in the Peninsular war will consist of solely British, or solely Portuguese, troops.

Early Peninsular War (1808-09) British & Allies

Elites

Up to half the units in the army may be of Elite status.

Divisions, Brigades & Batteries

Your army may include:

Type	Availability	Consisting of:
Infantry Division	Your army may include any number of infantry divisions. Each division may be either British or Portuguese, but cannot be mixed.	3-4 infantry units, and 0-1 rifle detachments, plus a divisional general. If British, one unit may be a Highlander or light infantry unit. One division in the army may include a single rifle regiment instead of light infantry/Highlanders. If Portuguese, one unit may be light infantry.
Foot Artillery Battery	Your army may include one foot artillery battery for every five infantry units in the army. If your army includes four units or less you may choose a half battery at half the points cost.	2 foot artillery guns & 1 howitzer.
Cavalry Brigade	Your army may include one cavalry brigade as long as it includes at least one infantry division.	0-1 British light cavalry unit and/or 0-1 Portuguese Cavalry unit, plus a divisional general (not a brigade commander).

Units

Choose the units that make up your divisions from the following:

Troop Type	Points Cost	Unit Size	Morale/Class
Rifle Detachment	10 points per detachment	3	Skirmishers
Portuguese Infantry	15 points per unit	6	Militia Infantry
British Line Infantry	20 points per unit	6	Line Infantry
Highlanders	25 points per unit	6	Line Infantry
Elite Highland	30 points per unit	6	Elite Infantry
Light Infantry	30 points per unit	6	Elite Infantry
Elite Line Infantry	25 points per unit	6	Elite Infantry
Rifle Regiment	50 points per unit	6	Guard Infantry
Light Cavalry	35 points per unit	6	Light Cavalry
Portuguese Cavalry	30 points per unit	6	Militia Cavalry
Foot Artillery Battery	60 points	2 horse artillery & 1 howitzer	Artillery

British Foot Artillery

Peninsular War (1810)
British & Allies

Elites

Up to half the units in the army may be of Elite status.

Divisions, Brigades & Batteries

Your army may include:

Type	Availability	Consisting of:
Infantry Division	Your army may include any number of infantry divisions. Each division may be either British or Portuguese, but cannot be mixed.	3-4 infantry units, and 0-1 rifle detachments, plus a divisional general. If British, one unit may be a Highlander unit. If Portuguese, one unit may be Cacadores.
Light Infantry Division	Your army may include one light infantry division. Each division may be either British or Portuguese, but cannot be mixed.	3-5 light infantry units, 1-4 rifle detachments (4-9 units in total), plus a divisional general. One rifle regiment may be chosen in place of a light infantry unit.
Foot Artillery Battery	Your army may include one foot artillery battery for every five infantry units in the army. If your army includes four units or less you may choose a half battery at half the points cost.	2 foot artillery guns & 1 howitzer.
British Cavalry Brigade	Your army may include one British cavalry brigade.	One unit of British cavalry (either light or heavy), plus a divisional general (not a brigade commander).
Portuguese Cavalry Brigade	Your army may include one Portuguese cavalry brigade.	One unit of Portuguese cavalry, plus a brigade commander.

Units

Choose the units that make up your divisions from the following:

Troop Type	Points Cost	Unit Size	Morale/Class
Rifle Detachment	10 points per detachment	3	Skirmishers
Portuguese Infantry	15 points per unit	6	Militia Infantry
British Line Infantry	20 points per unit	6	Line Infantry
Highlanders	25 points per unit	6	Line Infantry
Elite Highland	30 points per unit	6	Elite Infantry
Light Infantry	30 points per unit	6	Elite Infantry
Portuguese Cacadores	20 points per unit	6	Militia Infantry
Elite Line Infantry	25 points per unit	6	Elite Infantry
Rifle Regiment	50 points per unit	6	Guard Infantry
Light Cavalry	35 points per unit	6	Light Cavalry
Portuguese Cavalry	30 points per unit	6	Militia Cavalry
Foot Artillery Battery	60 points	2 horse artillery & 1 howitzer	Artillery

British Infantry Officer.

ALLIED PENINSULAR ARMIES (1811-12)

PORTUGUESE UNITS

By this point, the Portuguese army had undergone extensive retraining. Portuguese troops have the same abilities and special rules as British troops. Portuguese Cacadores count as British rifles and up to half the unit can detach as skirmishers.

ELITES

Up to half the units in the army may be of Elite status.

DIVISIONS, BRIGADES & BATTERIES

Your army may include:

Type	Availability	Consisting of:
Infantry Division	Your army may include any number of infantry divisions. Each division may be either British or Portuguese, but cannot be mixed.	3-4 infantry units, and 0-1 rifle detachments, plus a divisional general. If British, one unit may be a Highlander unit. If Portuguese, one unit may be Cacadores.
Light Infantry Division	Your army may include one light infantry division. Each division may be either British or Portuguese, but cannot be mixed.	3-5 light infantry units, 1-4 rifle detachments (4-9 units in total), plus a divisional general. One rifle regiment may be chosen in place of a light infantry unit.
Foot Artillery Battery	Your army may include one foot artillery battery for every five infantry units in the army. If your army includes four units or less you may choose a half battery at half the points cost.	2 foot artillery guns & 1 howitzer.
British Cavalry Brigade	Your army may include one British cavalry brigade.	1-2 units of British cavalry (either light or heavy), plus a divisional general (not a brigade commander). Both units must be of the same type.
Portuguese Cavalry Brigade	Your army may include one Portuguese cavalry brigade.	One unit of Portuguese cavalry, plus a brigade commander.

British Army.

British Infantry Officer.

Units

Choose the units that make up your divisions from the following:

Troop Type	Points Cost	Unit Size	Morale/Class
Rifle Detachment	10 points per detachment	3	Skirmishers
British Line Infantry	20 points per unit	6	Line Infantry
Highlanders	25 points per unit	6	Line Infantry
Elite Highland	30 points per unit	6	Elite Infantry
Light Infantry	30 points per unit	6	Elite Infantry
Elite Line Infantry	25 points per unit	6	Elite Infantry
Portuguese Militia	15 points per unit	6	Militia Infantry
Portuguese Infantry	20 points per unit	6	Line Infantry
Portuguese Cacadores	25 points per unit	6	Light Infantry
Rifle Regiment	50 points per unit	6	Guard Infantry
Light Cavalry	35 points per unit	6	Light Cavalry
Heavy Cavalry	40 points per unit	8	Heavy Cavalry
King's German Legion Dragoons	45 points per unit	8	Heavy Cavalry
Portuguese Cavalry	30 points per unit	6	Militia Cavalry
Foot Artillery Battery	60 points	2 horse artillery & 1 howitzer	Artillery

The Iberian Peninsula c. 1812.

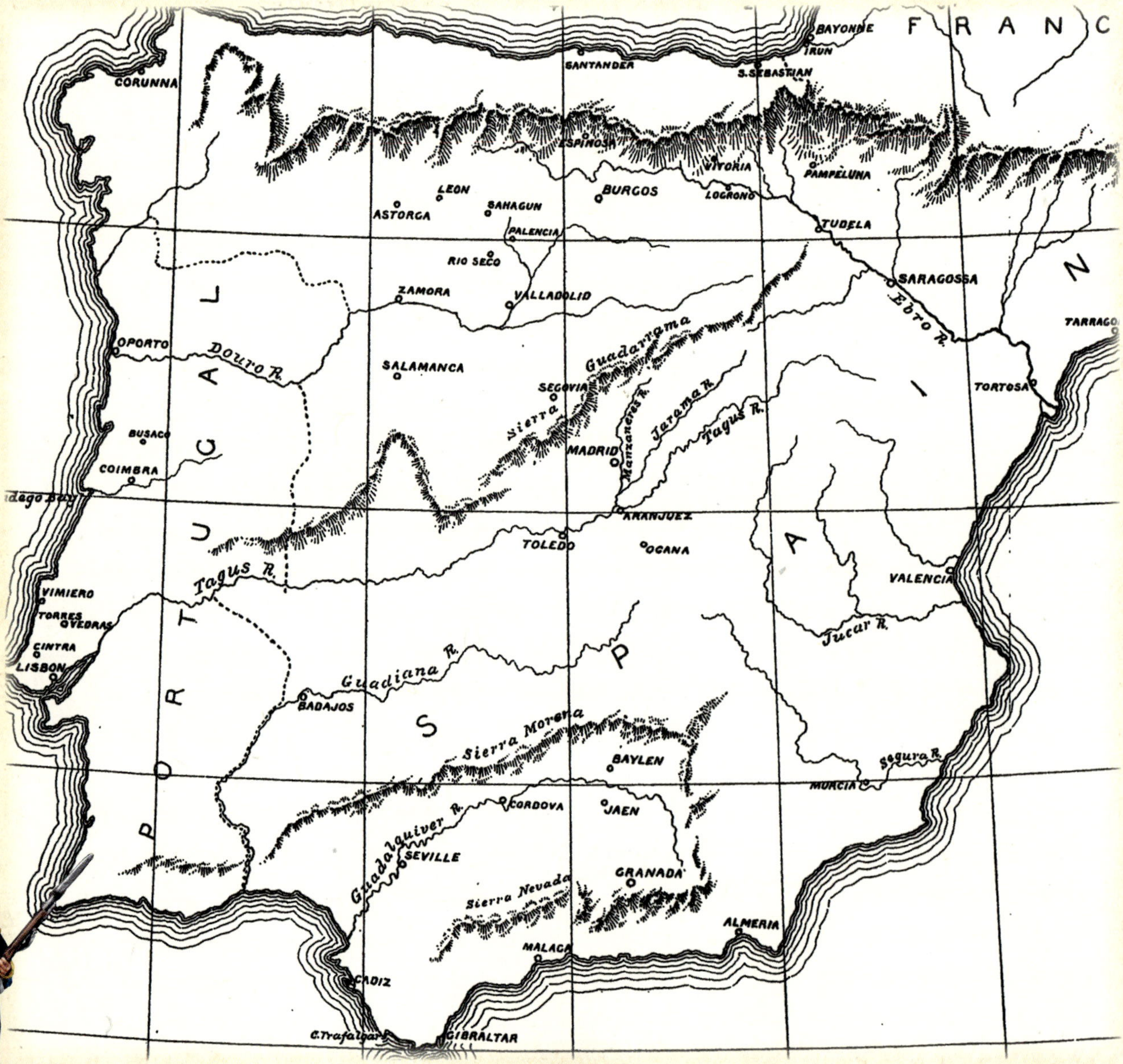

Portuguese Infantryman.

LATE ALLIED PENINSULAR ARMIES (1813-14)

PORTUGUESE UNITS

By this points, the Portuguese army had undergone extensive retraining. Portuguese troops have the same abilities and special rules as British troops. Portuguese Cacadores count as British rifles and up to half the unit can detach as skirmishers.

ELITES

Any number of units infantry may be of Elite status; all of them, if you wish.

DIVISIONS, BRIGADES & BATTERIES

Your army may include:

Type	Availability	Consisting of:
Infantry Division	Your army may include any number of infantry divisions. Each division may be either British or Portuguese, but cannot be mixed.	3-4 infantry units, and 0-1 rifle detachments, plus a divisional general. If British, one unit may be a Highlander unit. If Portuguese, one unit may be Cacadores.
Light Infantry Division	Your army may include one light infantry division. Each division may be either British or Portuguese, but cannot be mixed.	3-5 light infantry units, 1-4 rifle detachments (4-9 units in total), plus a divisional general. One rifle regiment may be chosen in place of a light infantry unit.
Guards Division	Your army may include one guards division if it includes at least two infantry divisions.	2-4 Guard infantry units, plus a divisional general.
Artillery Battery	Your army may include one foot or horse artillery battery for every four infantry units in the army. If your army includes three units or less you may choose a half battery at half the points cost.	Foot Artillery Battery consists of 2 foot artillery guns & 1 howitzer. Horse Artillery Battery consists of 2 horse artillery guns & 1 howitzer.
British Cavalry Brigade	Your army may include one British cavalry brigade for every two infantry divisions in the army. One brigade may be heavy, the others must be light.	1-2 units of British cavalry (either light or heavy), plus a divisional general (not a brigade commander). Both units must be of the same type.
Portuguese Cavalry Brigade	Your army may include one Portuguese cavalry brigade.	One unit of Portuguese cavalry, plus a brigade commander.
Household Cavalry Brigade	Your army may include one household cavalry brigade if it includes at least two cavalry brigades.	2-3 Guard heavy cavalry units, plus a brigade commander.

British Heavy Cavalry.

UNITS

Choose the units that make up your divisions from the following:

Troop Type	Points Cost	Unit Size	Morale/Class
Rifle Detachment	10 points per detachment	3	Skirmishers
British Line Infantry	20 points per unit	6	Line Infantry
Highlanders	25 points per unit	6	Line Infantry
Elite Highland	30 points per unit	6	Elite Infantry
Light Infantry	30 points per unit	6	Elite Infantry
Elite Line Infantry	25 points per unit	6	Elite Infantry
Portuguese Militia	15 points per unit	6	Militia Infantry
Portuguese Cacadores	25 points per unit	6	Light Infantry
Rifle Regiment	50 points per unit	6	Guard Infantry
Foot Guards	40 points per unit	6	Guard infantry
Light Cavalry	35 points per unit	6	Light Cavalry
Heavy Cavalry	40 points per unit	8	Heavy Cavalry
King's German Legion Dragoons	45 points per unit	8	Heavy Cavalry
Guard Cavalry	50 points per unit	8	Guard Heavy Cavalry
Portuguese Cavalry	30 points per unit	6	Militia Cavalry
Foot Artillery Battery	60 points	2 foot artillery & 1 howitzer	Artillery
Horse Artillery Battery	75 points	2 horse artillery & 1 howitzer	Artillery
Rocket Battery	60 points	2 rockets & 1 horse artillery	Artillery

British Hussars and Heavy Dragoons.

THE AUSTRIAN CAMPAIGN

While Napoleon was in Spain, and with his armies in central Europe depleted, Britain encouraged Austria to again take the field, with the promise of a secondary British effort at Walcheren. The Austrians struck on two fronts: one army under the Archduke Charles invaded Bavaria while a second, under the Archduke John, struck at Italy. The French were caught off guard. Marshal Berthier, left in charge in Germany, had disposed of his forces poorly and was facing defeat, while Eugene, Viceroy of Italy, was defeated at Sacile.

Faced with this threat to his German and Italian allies, Napoleon left Spain and reorganised his forces in Bavaria. He immediately went on the offensive, crossing the Danube and again pursuing his favourite tactic; he split the Austrian forces at the battle of Abensberg on the 19th/20th April before attacking each section in turn.

He first went for the left wing under General Hiller, while Davoust held the remaining force. Napoleon's attempt to encircle the left wing was foiled the next day at the battle of Landshut. But, undeterred, he switched his attack to the Austrian right-wing under Archduke Charles. Charles attacked Davoust's army at Eckmuhl on the 22nd April, but Napoleon's arrival swung the battle. His men were now tired after three days of continuous marching and battles and Napoleon was unable to pursue his foe and so gain the overwhelming victory he sought. Charles slipped away but covered his retreat by garrisoning the town of Ratisbon which fell on the 23rd. Napoleon entered Vienna on the 13th May having encountered no further resistance.

THE BATTLE OF ASPERN-ESSLING

The river Danube flowed to the east of the capital and the Archduke Charles formed his forces up on the far side of the river. Napoleon, using the island of Lobau in the Danube as a bridging point, established a bridge-head at the villages of Aspern and Essling. For two days over the 21st and 22nd May the French forces were subjected to heavy attack and, although the French held their ground, the position was becoming untenable as insufficient reinforcements could be brought over the river. Napoleon took the decision to withdraw.

Napoleon redrew his plans and set about overcoming earlier mistakes. Meanwhile, Eugene pushed the Austrians out of Italy by defeating Archduke John at Raab on the 16th June before moving to support Napoleon.

THE BATTLE OF WAGRAM

By the 4th July Napoleon, who had built up some of his forces on the island of Lobau, started a full scale crossing with one hundred and ninety thousand men – they were largely unopposed and allowed set up their bridgehead. His army was composed of three corps under Massena, Bernadotte and Oudinot, while the Imperial Guard cavalry units and two Italian corps were held in reserve – the latter under Eugene and MacDonald. Archduke Charles had positioned his one hundred and forty thousand men on the Heights of Russbach.

The next day Napoleon deployed his army around Aspern

Austrians face the French.

and Essling. The Austrian outpost divisions were pushed back so that by noon all of the area was in French hands. Despite night falling, Napoleon ordered his troops to attack. The initial attack by MacDonald's troops was poorly coordinated and although it took the high ground beyond the town of Wagram, heavy fire from the Austrians forced it to retire.

Fighting ceased until the following day when, at dawn, the Austrians took the offensive with a feint towards the French right wing. The real attack was then launched against the village of Aderklaa on the French left where two Austrians corps succeeded in throwing back Bernadotte's Saxons. Napoleon countered with a Grand Battery of one hundred and twelve cannon which decimated the Austrian forces. Following this, Massena's corps was sent to stabilize the situation.

Oudinot and Davoust on the right flank had successfully held the initial Austrian attack and were now pushing them back. Napoleon seized the opportunity and ordered MacDonald's corps against the centre of the advancing Austrians. Forming eight thousand men into a wedge and after ferocious hand-to-hand fighting, the French broke through the Austrian forces. The battle was won.

With its capital in French hands, its army defeated and no sign of any support from the other powers, Austria had no choice but to sue for peace. She had suffered a humiliating defeat. Napoleon extracted large areas of land and brought Austria into his economic war against Britain.

The battle had not been without its cost to the French; she had incurred a further thirty-four thousand casualties in addition to the twenty thousand suffered previously at Aspern-Essling. France would find it difficult to replace these soldiers in the coming years. Although unrecognised at the time, Wagram was to be seen as a turning point of the war. Napoleon's invincibility had been shaken at Aspern-Essling and the difficulty of extracting the final victory was both indicative of the gradual decline in the quality of his troops and the increasing experience and competence of his opponents who had learned from their previous errors.

Polish Grenadier (Knötel).

French Voltigeur- (Knötel).

Napoleon and the Pope.

AUSTRIAN ARMIES

UNIFORMS AND COLOURS

EARLY GERMAN INFANTRY

Throughout the Napoleonic Wars, the mainstay of the Austrian army was its German fusiliers; these men with the helmet issued in 1798 are dressed for the period from Marengo to Wagram. The traditional white coat of the Austrian infantry is decorated with an incredible variety of facing colours: dark red, Imperial yellow, sky blue, dark blue, black, poppy red, grass green, apple green, parrot green (lots of them flying around the Tyrol!), pink, dark brown, madder red, violet, sulphur yellow, light brown, crab red, sea green, light pike grey, light blue, rose, carmine, orange-yellow, steel green and pale red (mauve); the range of facing colours for the Austrians means a colourful order of battle, with the likelihood of no repetition.

Unit of early Austrian "German" Infantry.

EARLY HUNGARIAN INFANTRY

The Hungarian infantry of the Austrian army were distinguished from their 'German' counterparts by slight variations in the style of coat, sky blue trousers decorated with yellow and black braid, and ankle length boots.

Early Austrian "Hungarian" Infantry.

LATE GERMAN INFANTRY

In 1806 the Austrian infantry adopted a shako to replace the 1798 helmet. However an irregular supply meant that even as late as 1809 both types of headgear were still in service. Those interested in recreating the Austrian army of the 1809 campaign should be aware that priority had been given to replacing the helmets of the Hungarian infantry first, so a mixture of styles may still be apparent amongst their German counterparts.

Late Austrian "German" Infantry.

LATE HUNGARIAN INFANTRY

The major differences between the German and Hungarian infantry remained the same. Hungarian infantry received the 1806 style shako sooner than the Germans, so units still displaying the earlier 1798 headgear will be much less in evidence.

Late Austrian "Hungarian" Infantry.

German Grenadiers

Both German and Hungarian Grenadiers wore a bearskin cap of a unique design. It had a leather peak below a large brass plate with a yellow and black pom-pom on the right side. The rear of the cap gave the impression of having been cut away; from the top of the cut away section a bag hung, in facing colour decorated with lace (in 1811, all bags were ordered to be yellow with white lace, although this was probably not fully implemented until 1815).

Grenadiers companies from one regiment were often placed alongside grenadier companies from another regiment to form a crack unit. Such units therefore often contained two different facing colours.

Unit of Austrian "German" Grenadiers.

Hungarian Grenadiers

As with the line infantry, grenadier units were divided into those of German and Hungarian origin. The differences in uniform remained much the same.

Austrian Jaegers

As in other armies, the jaegers were light infantry, their name meaning hunter. Like the Grenadiers, the Jaeger wore a headdress of an original Austrian style, with a tall, flat crown and wide brim the hat was turned up on the left side and carried a yellow and black pom-pom on the front top edge of the crown. The uniform was grey faced dark green and the equipment black.

Austrian Grenz Infantry

Grenzers were troops from the Austrian Empire's Balkan frontier, who initially fought as light infantry but were not the force they had been during the wars of Frederick the Great and Maria Theresa. By the time of the Napoleonic Wars the Grenz infantry still numbered a considerable body of men, even if attempts to turn them into regular infantry had damaged their effectiveness as light infantry. The orders of dress for the Grenz regiments is confusing but it seems likely they fought in the 1805 campaign in white uniforms and in brown in 1809, the facing for both were: Emperor yellow, orange-yellow, crab red, pale red, light pike grey, crimson, sky blue, pink and parrot green.

> *Unity of command is essential to the economy of time. Warfare in the field was like a siege: by directing all one's force to a single point a breach might be made, and the equilibrium of opposition destroyed.*
> — Napoleon Bonaparte

Austrian "Hungarian" Grenadiers.

Austrian Jaegers.

Austrian Grenz Infantry.

Early Austrian Landwehr

The Landwehr where the militia of the Austrian army. The Austrian Landwehr were formed in time to take part in the campaign of 1809, though their contribution was negligible. Hungary refused to take part in the scheme so its implementation was restricted to the 'German' lands. The organisation was divided into 'normal' and 'volunteer' units, the latter being by far the better motivated troops. It was further divided into areas called directorates, there were three of these: Bohemia and Moravia, Inner Austria, and Upper and Lower Austria. Standardisation in appearance was rarely achieved and in most units only officers and NCOs wore the uniform, (others had to be content with part of the ensemble or even just a cockcade). Of the units available in this range, the Inner Austrian wore grey coats over their own clothes with white collars, while the Bohemians were dressed in brown jackets faced and braided red, blue breeches and black equipment.

Unit of Austrian Landwehr Infantry.

Late Austrian Landwehr

In the year following the defeat of 1809 the militia were banned, as part of the peace terms. However when Austria renewed hostilities they were able to raise new units and made a serious attempt to dress them in a uniform manner. As each battalion was now linked to a regular unit they were to wear a grey coat with the facing colours of their parent unit.

Late Austrian Landwehr.

Austrian Artillery

The Austrian Artillery had slipped somewhat from it's position of pre-eminence, during the eighteenth century. The personnel were dressed in orange-brown coats, with red facings and the artillery equipment was painted yellow.

Small Austrian Artillery Battery (left) prepares to take on the much larger French Grand Battery.

Austrian Dragoons/ Chevauxlegers

In 1798 the regiments of Dragoons and Chevauxlegers were transformed into fifteen regiments of Light Dragoons. Unfortunately this was the first in a series of changes that would not be resolved till the end of the wars. Three units were then disbanded and the remaining twelve were again reformed, in 1801, into six regiments of Dragoons and six of Chevauxlegers.

As Light Dragoons the regiments had worn green coats but with the 1801 reorganisation the Dragoons changed to white, the Chevauxlegers retaining the green. All this was attended with the proviso that existing uniforms should be allowed to wear out, this meant that as late as Austerlitz certain Dragoon regiments were still in green. Meanwhile some of the Chevauxlegers had also adopted white, which was confirmed in 1805 when it was agreed that those Chevauxlegers now wearing white should retain it.

Dragoons: All white coats and white metal buttons. Facing colours: 1st black, 2nd dark blue, 3rd pompadour (dark) red (later bright red), 4th light red (later bright red), 5th dark green, 6th light blue.

Chevauxleger: All except the 2nd had yellow metal buttons. Coats and facing colours: 1st green coats/pompadour red facings, 2nd green coats/dark green facings, 3rd white coats/light red, 4th green coats/dark blue facings, 5th white coats/light blue facing, 6th white coats/black facings, 7th (raised 1814) green coats/crimson facings (white metal buttons). All Austrian cavalry had a shabraque of the same design – red with pointed rear corners and rounded front, edged with yellow and black lace and covered with a sheepskin over the saddle area.

Austrian Cuirassiers

The Austrian cavalry had a high reputation at the beginning of the Napoleonic Wars, none more so than the Cuirassiers. Although the reputation was tarnished a little by the lack of success of Austrian arms, the cavalry remained good. The cuirassiers wore white coats of the same style as the Dragoon and Chevauxleger over which was worn a front plate only. The facing colours of the regiment were: dark red, black, grass green, light blue, dark blue, and scarlet.

Austrian Cuirassiers.

Austrian Hussars

Hussars were light cavalry which had, in one form or another, been in use in various Eastern European armies since the Middle Ages. Their heyday came in the army of Frederick the Great and the Austrian Hussar of the Napoleonic Wars was as close to the original as any nation could manage. Twelve, later eleven, regiments were dressed in matching dolman and pelisse with breeches in red or to match the upper garments, grey overalls were worn on campaign. The following colours were used for the dolman and pelisse combination: dark blue, light blue, parrot green, dark green, pike grey. To add to the ensemble the shakos were coloured: black, red, grey, bright blue and grass green. Hussars did not carry standards on campaign.

Austrian Napoleonic Uhlans

Uhlans (named for an earlier light cavalry type) were the lancers of the Austrian army. There were three units of Austrian Uhlans, with a fourth raised in 1814. They were all dressed in green faced red with the distinguishing characteristic being the colour of the czapka: 1st red, 2nd dark green, 3rd scarlet and 4th white.

Austrian Landwehr Officer

Austrian Army (1809) and the Army of Bohemia (1813-14)

Austria was Napoleon's most irrepressible of enemies, frequently vanquished but always able to eventually return to the fray. The Austrian army was a polyglot force reflecting the various peoples within the Empire: regiments were divided into 'German' and 'Hungarian', but contained Austrians, Poles, Italians and Dutch in the former, and Hungarians, Croatians and Romanians in the latter. The officers were principally Austrian, largely from the lesser nobility. Training was inconsistent and the need to recruit large numbers in wartime led to many untrained men in each battalion. This, together with government cost-cutting, conflicting orders and frequent reorganisations, resulted in a less than efficient force in the field.

Austrian Special Rules

The following special rules apply to Austrian Armies, and Bohemian armies of 1813-14:

German & Hungarian Troops

The army was massive and split along geographical lines; German fusilier infantry were organised into regiments containing a number of very large battalions of six companies, and each company had over two hundred men, Hungarian units were even larger!

To represent this on the tabletop Austrian line units should be larger than the typical unit size of 24 models. German line infantry (fusilier) units should have 32 models each, while Hungarian units should have 40 models! This affects their firing and morale as follows:

German Units

All Germany line infantry units in line formation roll 8D6 instead of 6D6 when firing volleys and 3D6 dice in close combat for each unit it is engaged with, up to a total of 8D6. The morale is also increased to 3D6 dice; however when the unit has lost command and is attempting to regain it, roll 3D6 and discards the lowest dice.

Hungarian Units

When in Line formation, Hungarian units increase their number of firing dice to 10D6 when in line formation and roll 3D6 dice when in close combat for each unit it is engaged with, up to a total of 10D6 dice.

These increased unit sizes do not apply to militia, or to some specialised troop types in the army.

Grenadiers

Each infantry regiment had two grenadier companies, which were normally deployed with two other grenadier companies from another unit. With no guard troops in the army these grenadiers were often joined with others of their kind to form crack divisions.

Grenadiers are represented by the usual six stands. All grenadier units are of elite status.

Austrians engage the French.

Grenadiers were formidable at close quarters, so they count as Guards and receive +1 to hit in close combat.

Grenzers & Frei-Corps

While they were not of the same quality as in the previous century, Austrian armies still used Grenzers and Frei-Corps units to perform the role of light infantry. Units of Grenzers are represented by 24 models, with four models per base. You will need extra models (or some models based singly) to represent skirmishers if you are using the Advanced Rules.

Grenzers count as militia, but their volley fire is counted as being of Guard status and they receive an extra 1D6. They can deploy two stands of skirmishers.

Austrian Cavalry

Austrian cavalry was of excellent quality; their Hussars were the original embodiment of light cavalry and by the time of the Napoleonic Wars were more regular and highly organised than they had been for over 100 years.

Cuirassiers

Cuirassiers only wear a breast plate (with no back plate) and so cannot ignore further casualties after losing combat.

French General of a Division (Knötel).

Austrian Army (1809) and the Army of Bohemia (1813-14)

Divisions, Brigades & Batteries

Your army may include:

Type	Availability	Consisting of:
Infantry Division	Your army may include any number of infantry divisions.	2-5 infantry units, and 0-1 detachment of Jaegers, plus a divisional general. Up to two units may be militia. If your army contains only a single infantry division, one unit may be grenadiers and one unit may be grenzers, otherwise all units must be fusiliers, Hungarian line infantry or militia. No individual unit type may exceed the number of fusilier units in the division.
Grenadier Division	Your army may include one grenadier division for every two infantry divisions it contains.	2-4 grenadier units, but plus a divisional general.
Advance Guard Division	Your army may include one advance guard division.	1-2 light cavalry units, 1 light infantry unit (of your choice), 1 Jaeger detachment, 1-2 Grenzer units and 1 horse artillery battery.
Light Cavalry Brigade	Your army may include one light cavalry brigade for every infantry division it contains.	1-2 light cavalry units, plus a brigade commander; one of these may be lancers.
Light Cavalry Division	Two light cavalry brigades may be combined to form a division; you must meet the requirements to field both brigades.	As brigades; one brigade commander is promoted to divisional general.
Dragoon Brigade	Your army may include a single dragoon brigade if it contains at least three infantry divisions.	1-2 dragoon units, plus a brigade commander.
Cuirassier Division	Your army may include one cuirassier division for every grenadier division it contains.	Two brigades, each consisting of 1-2 cuirassier or dragoon unit. One brigade has a brigade commander, the other has a divisional general. The division may also include a horse artillery battery.
Foot Artillery Battery	Your army may include one foot artillery battery for every four infantry units (not counting Jaegers) it contains, up to a maximum of one battery per infantry division. If your army contains three infantry units or less, you may include a half battery at half the points cost.	3 foot artillery guns & 1 howitzer. Half battery consists of 2 foot artillery guns.
Heavy Artillery Battery	Your army may include one heavy artillery battery for every grenadier division of three or more units it contains.	2 heavy artillery guns & 1 howitzer.

There are in Europe many good generals, but they see too many things at once. I see one thing, namely the enemy's main body. I try to crush it, confident that secondary matters will then settle themselves. .

— Napoleon Bonaparte

Austrian Officer.

UNITS

Choose the units that make up your divisions from the following:

Troop Type	Points Cost	Unit Size	Morale/Class
Fusiliers	25 points per unit	8	Line Infantry
Grenadiers	30 points per unit	6	Elite Infantry
Hungarian Line Infantry	30 points per unit	10	Line Infantry
Grenzers	20 points per unit	6	Militia Infantry
Militia	20 points per unit	6	Militia Infantry
Hussars	35 points per unit	6	Light Cavalry
Lancers	40 points per unit	6	Light Cavalry
Chevaux-Legers	40 points per unit	6	Light Cavalry
Dragoons	40 points per unit	8	Heavy Cavalry
Cuirassiers	45 points per unit	8	Heavy Cavalry
Jaegers	10 points per detachment	3	Skirmishers
Foot Artillery	60 points	3 foot & 1 howitzer	Artillery
Heavy Artillery	60 points	2 heavy & 1 howitzer	Artillery
Horse Artillery	60 points	3 horse	Artillery

The Battle of Ratisbon, 23 April, 1809. The pursuit of the Austrians after the French victory at Eckmuhl.

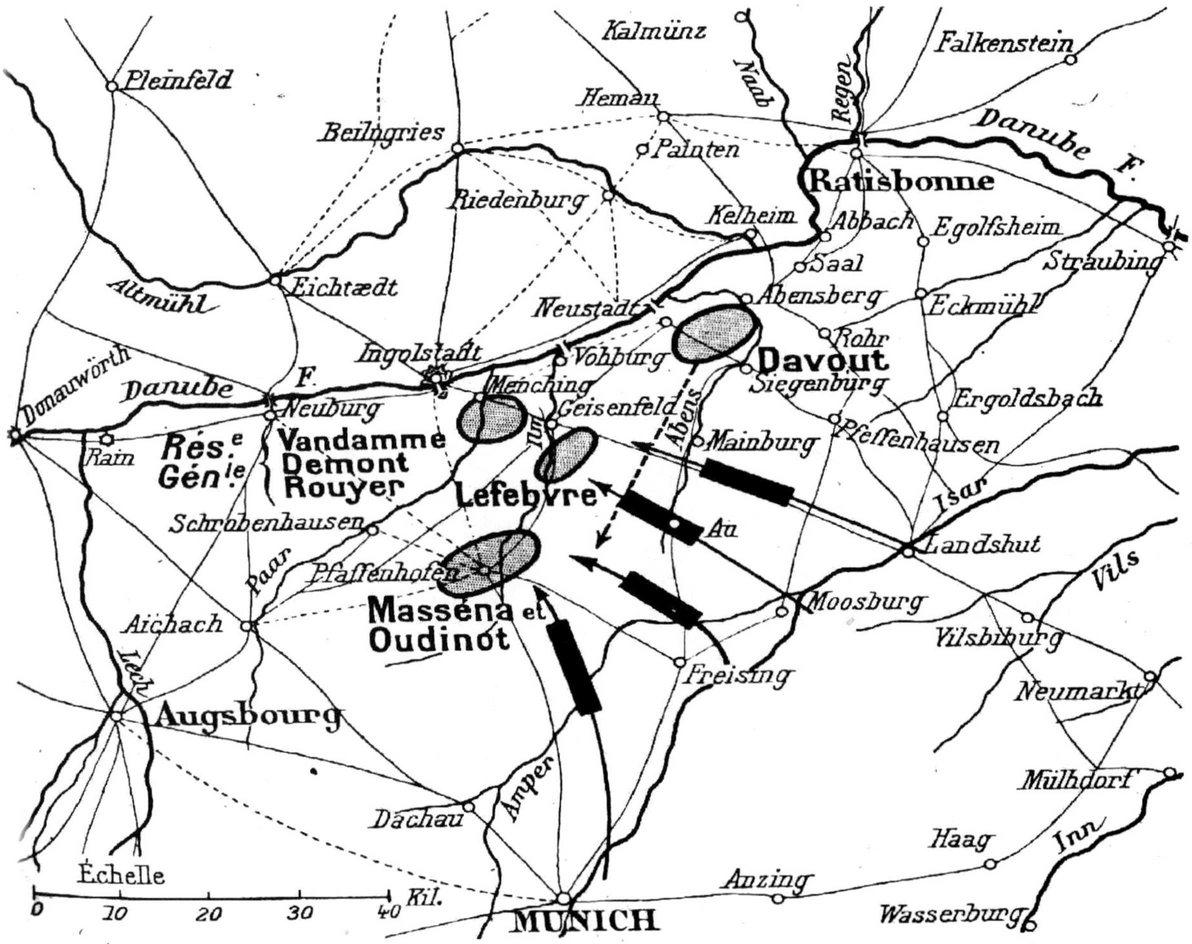

C'est alors le corps Davout qui tombe dans le flanc droit de l'ennemi occupé à combattre les corps Lefebvre, Oudinot et Masséna.

Austrian "German" Grenadier

Napoleon's Army in Austria (1809)

To counter Austria's renewed antagonism towards France, Napoleon left Spain and headed off to fight his old Austrian adversary. Until the Emperor arrived Marshal Davoust and his army were all that stood in opposition to the Austrian forces. After years of glory the quality of the French army was starting to deteriorate. To give his forces more backbone Napoleon experimented with the reintroduction of battalion guns (he had hundreds of captured Prussian guns to fill this role). The newly formed Young Guard was also deployed for the first time and the reliance on allied, mostly German, troops was to change the face of the army.

All French rules as given in the Peninsular Wars list apply to this army (see page 131).

Bavarian or Wurtemburg troops (use the Bavarian list) can directly replace any infantry or cavalry divisions.

Carabineers did not wear armour in this campaign so they cannot ignore any hits in close combat.

The Austrian Campaign (1809) French

Divisions, Brigades & Batteries

Your army may include:

Type	Availability	Consisting of:
Infantry Division	Your army may include any number of infantry divisions.	3-5 infantry units, plus a divisional general; up to two units may be light infantry. Each line infantry unit may have an attached battalion gun. Battalion gun +10 points per unit.
Cavalry Brigade	Your army may include one cavalry brigade for each infantry division in the army.	1-2 units of cavalry; both units must be of the same type, except for cuirassiers and carabineer units which may be combined. At least half the cavalry units in the army must be light cavalry.
Cavalry Division	Two cavalry brigades may be combined into a cavalry division. You must meet the requirements for fielding both brigades.	As brigades; both brigades must be of the same type, with the exception of cuirassiers and carabineer units which may be mixed within a division. One brigade commander is promoted to divisional general.
Artillery Battery	Your army may include one artillery battery for every four infantry units it contains. If your army contains three infantry units or less, you may include a half battery at half the points cost.	The first artillery battery in your army must be a foot artillery battery, consisting of 3 foot artillery guns & 1 howitzer. Further artillery batteries may be either foot artillery batteries, or heavy artillery batteries consisting of 3 heavy artillery guns & 1 howitzer.
Horse Artillery Battery	Your army may include one horse artillery battery for every cavalry division it contains, or four every four units of cavalry which are not part of a division.	3 horse artillery guns & 1 howitzer.
Young Guard Division	Your army may include a single Young Guard division if it contains at least two infantry divisions.	3-4 Young Guard infantry units; up to half the units may be Young Guard light infantry.
Guard Division	Your army may include a single guard division if it contains at least three infantry divisions.	2-4 Old Guard infantry units, plus an Old Guard divisional general.

Guard Cavalry Brigade	Your army may include a single guard cavalry brigade if it includes a guard division or at least three infantry divisions. If your army includes two or more guard divisions, you may include two guard cavalry brigades.	The brigade may be either light or heavy. Guard light cavalry brigades consists of 1-3 units of guard light cavalry; guard heavy cavalry brigades consist of 1-2 units of guard heavy cavalry, plus a divisional general (not a brigade commander) in both cases.
Guard Artillery	You may upgrade artillery and horse artillery batteries to guard artillery if you army includes a guard infantry or cavalry division.	This is an upgrade and does not bring your army any additional units: any number of artillery batteries or horse artillery batteries can be upgraded to guard status.

UNITS

Choose the units that make up your divisions from the following:

Troop Type	Points Cost	Unit Size	Morale/Class
Line Infantry	20 points per unit	6	Line Infantry
Light Infantry	25 points per unit	6	Line Infantry
Young Guard Infantry	25 points per unit	6	Elite Infantry
Young Guard Light Infantry	30 points per unit	6	Elite Light Infantry
Old Guard Infantry	40 points per unit	6	Old Guard Infantry
Light Cavalry	35 points per unit	6	Light Cavalry
Dragoons	40 points per unit	8	Heavy Cavalry
Cuirassiers	45 points per unit	8	Heavy Cavalry
Carabineers	45 points per unit	8	Heavy Cavalry
Guard Light Cavalry	50 points per unit	6	Guard Light Cavalry
Guard Heavy Cavalry	55 points per unit	8	Guard Heavy Cavalry
Foot Artillery Battery	60 points/30 points	3 foot artillery & 1 howitzer	Foot Artillery
Heavy Artillery Battery	75 points	3 heavy artillery & 1 howitzer	Heavy Artillery
Horse Artillery Battery	75 points	2 horse artillery & 1 howitzer	Horse Artillery
Guard Foot Artillery	+10 points	-	Guard Foot Artillery
Guard Heavy Artillery	+10 points	-	Guard Heavy Artillery
Guard Horse Artillery	+10 points	-	Guard Horse Artillery

French on campaign.

It is the business of cavalry to follow up the victory, and to prevent the beaten army from rallying.

— Napoleon Bonaparte

French Colonel of Cuirassiers (Knötel).

Russian Campaign 1812 The Patriotic War

Following two years of peace in most of Europe, Napoleon's hold over the continent seemed almost absolute; only Britain of the main powers remained at war with France. All was not well in the Empire, however; cracks were beginning to show. The strains imposed by Napoleon's Continental System and the growing resentment at French rule caused stirrings of unrest. By 1812 Russia, angered by the creation of a Polish state on her boundary and threatened with economic ruin by the cutting off of trade with Britain, moved to distance herself from France.

Napoleon responded by forming an army over half a million strong, drawing men from every corner of his empire and including Prussian and Austrian contingents. He made a show of strength in Germany by holding court in Dresden to which all the German rulers were invited in an attempt to bring the Russian Emperor, Alexander, back into the fold. Napoleon knew that an invasion of Russia was a daunting task. The distances involved had proved disastrous for other armies in the past and he well knew the fighting qualities of the Russian soldiers from his previous campaigns.

The Russian army was poorly led and its soldiers were largely peasant conscripts, but what it lacked in military prowess it more than made up in manpower and stubbornness. While it could boast a European style army, many of its irregular forces such as the Cossacks and the tribes from the east and south were little changed from the armies of medieval times and some were even armed with bows and arrows.

Napoleon planned his campaign meticulously, ensuring that supplies would be available and stockpiles arranged at strategic centres. He also planned for a quick and decisive knockout blow, wishing to avoid at all costs being drawn into a long war of attrition or being sucked deeper into Russia. In response, the Russians assembled two armies, one under the German-born General Barclay de Tolly and the second under the Ukrainian General Bagration. Russia also quickly arrived at an armistice with her longstanding enemy Turkey, which freed up more forces under Admiral Chichagov in the south – something Napoleon had not counted upon.

With his political sabre rattling making no obvious impact, Napoleon moved his troops across the River Niemen on 24th June, 1812. In what was to become the archetypal manoeuvre of the campaign the Russians withdrew. Despite French attempts to encircle the Russian armies, misunderstandings and poor leadership let the Russians slip away. By 3rd August the two Russian armies had united at Smolensk. Barclay de Tolly took overall command, but he now faced growing hostility from the Russian hierarchy due to his continual avoidance of battle, criticism which was fuelled in part by a resentment of de Tolly's German ancestry.

Russian Infantry Officer.

Napoleon again unsuccessfully tried to outflank the Russians, storming and capturing Smolensk on the 17th August and winning a minor victory at Valutino on the 19th. The Russian response was to withdraw still further. Angered at Barclay de Tolly's apparent inaction, Alexander replaced him with the aging Russian hero Mikhail Kutuzov. The withdrawal however, continued. The hot summer, the poor foraging available to the French armies and its allies and the need to leave men to protect its growing supply lines, was weakening and reducing the fighting strength of Napoleon's army. The Austrians in the south under Schwarzenberg were doing little more than marking time, while the French and her Prussian allies under MacDonald were camped outside Riga.

The Battle of Borodino

Each day brought Napoleon closer to Moscow, the ancient capital and holy city of Russia. He believed its capture would be a major blow to Russian morale. Sixty miles from the city Kutuzov stopped and set up a defensive position around the village of Borodino straddling the main road to the city. Fortifications were thrown up including the Great Redoubt (earthworks) and the Russian army finally stood to receive the French.

Napoleon saw the chance for the major victory he needed. He addressed his soldiers with the words: 'Soldiers! Behold the field of battle you have so much desired! Henceforth victory depends on you; it is necessary to us; it will give us plenty, good quarters for the winter and a speedy return to your country…'

On 7th September the two armies clashed but whether through the rumoured illness or lethargy, or due to the lack of men and a realisation of the need to conserve his forces, Napoleon's battle plan lacked the imagination of earlier years. He threw his soldiers forward in a murderous frontal attack and the two armies battered away at each other for the whole day in one of the bloodiest battles of the war. General Bagration himself was mortally wounded. Napoleon, conscious of the fact that he had committed almost his entire force, refused to send in his Imperial Guard anticipating that it may yet be needed later in the campaign. By so doing he lost the opportunity of the overriding victory he needed. By the next morning the Russian forces had again drifted away, leaving the road to Moscow open.

Moscow

A week later Napoleon entered the city but his hopes of an end to the campaign were dashed. The city was all but empty. His approaches to Alexander to sue for peace were not answered and within days much of the city had been set ablaze. Napoleon thought about his next move; to stay in

Moscow over winter, to march on the capital St Petersburg or to fall back to Smolensk and resume the campaign the following year. With no likelihood of peace, the onset of winter fast approaching and with Moscow offering little refuge for his army (while the Russians were growing stronger and possibly encircling him), he finally took the fateful step to withdraw. He left the city via the south-west but on 24th October Eugene's advance guard met the Russians at Maroyaroslavets where it was held.

Fearing the destruction of his army Napoleon turned north and retraced his route along the high road by which he had approached Moscow. The Russian winter and the retreat – whilst being pursued by the Russian army and Cossack bands – along a route already stripped bare of provisions by peasantry eager for revenge, was a disaster for the remaining troops. With superhuman effort the commanders kept their forces together and, despite parts of the army having to fight their way through the Russian forces, eventually made it to Smolensk. Its hopes of food and rest were dashed, though, partly through poor organisation and partly through insufficient supplies. The retreat continued.

By the end of November the French reached the banks of the Berezina where the Russians were waiting to cut them off. A thaw had led to the river being in full flow rather than iced over. Napoleon misled the Russians into thinking he was going to cross further south and set his engineers to work building bridges. In two days most of what remained of his army crossed over, but thousands of soldiers and civilians were left on the Russian shore when the bridges finally collapsed. Those that had escaped continued their slow, painful retreat to the Niemen which they had crossed just six months before with such high hopes.

Napoleon, conscious that his absence from Paris was leading to instability at home and recognising that his work in leading the army was finished, left what remained of his force on the 8th December. His aura of invincibility had been shattered, although his defeat had as much to do with the Russian weather as with the fighting skills of the Russians themselves.

Russians armies were provided with huge artillery batteries.

Russian Cuirassier.

Russian Armies

How many things apparently impossible have nevertheless been performed by resolute men who had no alternative but death.

— Napoleon Bonaparte

Russian Line Infantry

Grenadiers

Pavlov Grenadiers

The Pavlov Grenadiers were an elite guard regiment.

Jagers

Although classed as light infantry, jagers were usually used in the infantry role, the Russian army was not known for its infantry skirmish tactics.

MILITIA

Lacking the trained soldiery of many other nations, the Russian army frequently conscripted large numbers of militia. Some were so poorly equipped they fought only with pikes – a weapon largely abandoned centuries earlier!

Unit of Militia.

Unit of Muscovite Militia.

Militia pikemen.

DRAGOONS

CUIRASSIERS

Russian Cuirassiers and Dragoons.

Hussars

Mounted Jagers

Cossacks

Russian Artillery

Russians advance from a walled town to drive back the Bavarian infantry.

Russian Flags

Russian flags are a complex subject. There seem to have been a number of 'issues' of flags, in 1797, 1800, 1803, 1806/7 and 1813, but the old patterns were not always withdrawn, so that some regiments carried older pattern flags right through the Napoleonic Wars. Precisely which regiments did so is more of a problem. Only in the case of the Battle of Borodino do we get a clear idea of who was still carrying old flags, and then only for the particular regiments recorded here.

Up to 1813 each infantry regiment carried both coloured and white flags; after that the white flags were discontinued.

Using The Printed Flags

The flags have been drawn in the size that we recommend you use for your regiments (the size in the photographs). They may appear over-large, but this is the ideal size for visibility on the tabletop. It also makes them easier to copy and cut out! If you like smaller flags, reduce them in the copying process. You could always download them from our website **www.wargamesfoundry.com**.

Once you have copied them, cut them out, leaving some white surround.

Fix them around a suitable pole with white PVA glue.

Allow this to soak well into the flag, then press the two sides together around the pole.

Leave to cure for a while, keeping it pressed flat under an old book.

Bend the flag whilst it's still flexible, to create folds as if it's blowing in the wind.

Once the flag is hard trim it with a sharp pair of scissors so that none of the surrounding white paper shows.

You can now varnish it for extra strength or even add some painted-on highlights.

For extra strength and *bendability* you can make a sort of sandwich, with a thin bit of aluminium foil between the two 'sides' of the flag before you stick it together.

As an alternative to colour copying you might like to visit our web site (www.wargamesfoundry.com) and download the flags from there. Here you will also find a 'dark' version of each flag, so that you can paint on the next brighter level of colour and highlight yourself after printing them out; this will give less of a 'printed' and more of a 'professional painter' look.

* Carried at Borodino 1812

*Pavlov Grenadiers Coloured 1797 Issue**

*Pavlov Grenadiers White 1797 Issue**

Little Russia Grenadiers Coloured 1797 Issue

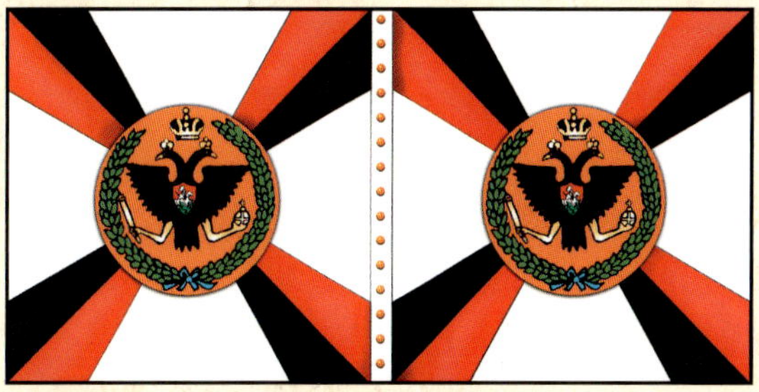

*Little Russia Grenadiers White 1797 Issue**

Guards Coloured 1800 Issue

Guards White 1800 Issue

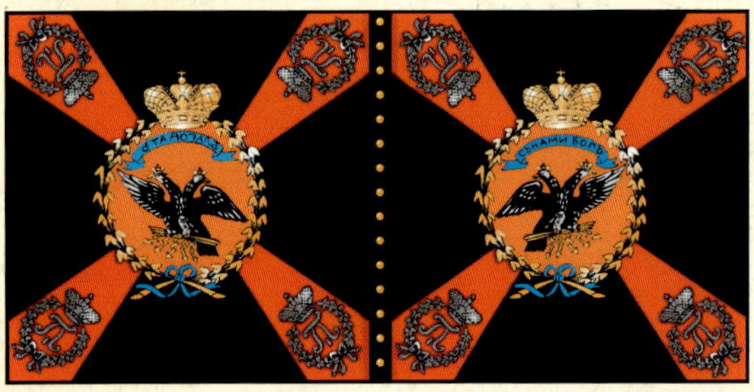

Brest Litovsk, Lithuania, Livonia, Smolensk Coloured 1800 Issue

Finland, Moscow Coloured 1800 Issue

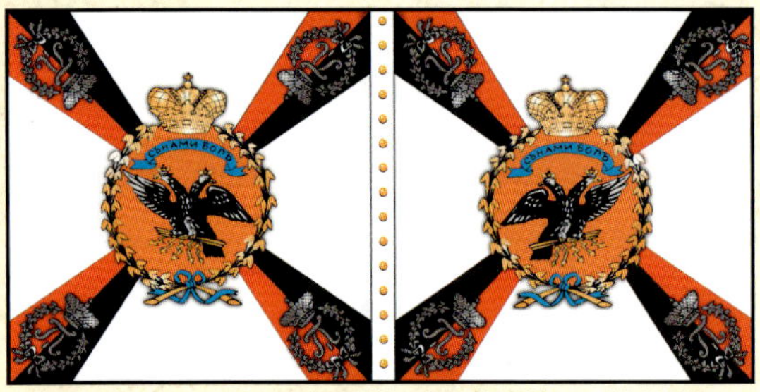

Brest Litovsk, Lithuania, Livonia, Smolensk White 1800 Issue

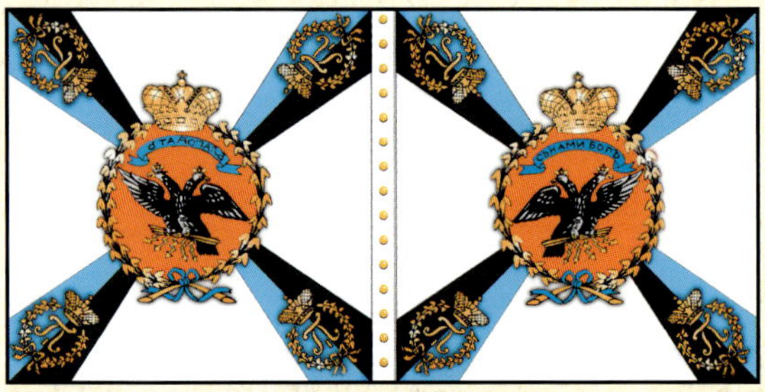

Finland, Moscow White 1800 Issue

Siberia Coloured 1803 Issue

Belostoksky MusketeersColoured 1807 Issue

White 1803 Issue

Azovsky Coloured 1807 Issue

Fanagorisky Grenadier Coloured St George 1807 Issue

All Line Infantry 1813 Issue

White St George 1807 Issue

White all Line Infantry 1813 Issue

Pavlov Grenadier Guards 1813 Issue

Finland Grenadier 1813 Issue

White St George 1813 Issue

Line St George 1813 Issue

Grenadier 1813 Issue

Isamailov Grenadier 1813 Issue

Crimea Coloured 1803 Issue

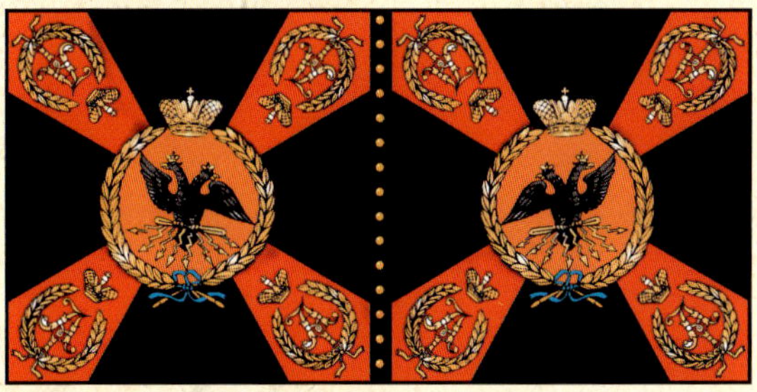

St Petersburg 1803 Issue

Kiev Coloured 1803 Issue

Ukraine Coloured 1803 Issue

Moscow Grenadier Coloured 1803 Issue

Permk Coloured 1807 Issue

Russian Uniforms

1. Regimental Drummer

Shown here wearing a fatigue cap (of the 1st battalion). Grenadiers' shakos used a tall red plume for drummers and musicians. Drumsticks were the same colours as the standard staff brown, yellow, black or white, according to the individual regiment.

2. Field Officer

Frockcoats were popular for most officer ranks in the field, together with the peaked version of the fatigue cap and grey, buttoned overalls. Gilt gorgets, (not always worn on campaign) and gold-fringed epaulettes were a sign of field officers. Officers of 'Jager' regiments would have dark green piped red cuffs and collars whilst Marines would have had dark green, piped white ones on their frockcoats.

Mounted officers' saddle cloths and holster covers were dark green with a gold double striped boarder (galloons) and red stripe between the gold. Jager saddlecloths were the same apart from a dark green stripe instead of red, and all gold galloons piped red. Marines were the same as Jagers but piped white instead of red.

3. Foot Artillery

The black facings piped red, typical of many a nations artillery would be repeated on the collars of their brown/grey greatcoat (shinel). 3(a) shows the fatigue cap. (b) Shows the guard lace (petlitze) on the collar and cuffs. The gun carriages were painted mid-green with black fittings. Barrels were bronze.

4. Marines

There were four Marine regiments serving in the army from 1812, three in the 25th Division and one in 28th Division as shown in the shoulder straps (b). The fatigue cap (a) has the same differences in piping as in the Infantry battalions, (but has green headband instead of red), likewise the sword knot for the companies. The shinel collar was dark green piped white.

5. Jagers

The figure shown is wearing the 1809 shako (kiwer) which continued in service alongside the 1812 kiwer with the concave top, although probably not in the same units. Carbiniers (one platoon in each battalion) had the same distinctions as grenadiers in the infantry (i.e. tall plumes and triple flamed grenade devices on kiwers and cartouche pouches). The brown-grey shinel had a dark green collar, piped red.

6. Infantryman (Musketeer)

The term 'musketeer' had been abolished in 1811 in favour of the term 'infantryman' and by the end of 1811 the regiments had lost their grenadier companies, which now formed grenadier divisions. The infantryman shown here is in regulation winter dress; leather booted overalls, with cords left attached to the shako, a common practice on campaign in the Russian army.

7. Grenadier

The figure shown here is wearing summer dress; white gaiter-trousers, which in winter were sometimes worn under the booted overalls. The 1809 kiwer sports the tall thin grenadier plume, which replaced the enormous 'bush' plume in 1811.

8. Infantryman in Shinel

This figure shows the typical campaign dress for summer or winter; fatigue cap and shinel (Greatcoat) which could be worn over a shirt, waistcoat, civilian frock-coat or smock, the latter two in winter. In 1811 cuffs of the greatcoat were ordered to be faced in colours of the Arms of Service i.e. red for infantry, dark green piped red for jager etc. This regulation does not seem to have been adhered to, although contemporary pictures of guards (probably Preobrazhenski or Litovski regiments) show them to have red cuffs, collars and shoulder straps on their greatcoats. The Infantryman here is from the 3rd battalion, 8th company of his regiment, shown by the light-blue piping on his fatigue cap and his sword knot. The 2nd battalion was usually retained as the depot battalion.

9. Line Officer

The officer illustrated wears the regimental summer dress used on campaign, including the knapsack worn by all but field officers. Officers were not always so neat on campaign, like the Infantry, items of civilian clothing were used, such as scarves and peasant smocks.

Epaulettes were the regimental strap colour, edged in gold lace with the number in gold. Gold fringes and gorgets were reserved for field officers. Gorgets were gilt with a silver eagle for captains, silver with gilt eagle and edge for 2nd captains, silver with gilt edge for lieutenants and silver for 2nd lieutenants. Officers are sometimes shown still in their old black bicorns with a black panache (hanging plume), Guards having a white panache together with red striped, grey overalls.

10. Pavlovski Guardsman 1813-15

After heroic fighting in the 1812 campaign the Pavlovski Regiment attained guard status and joined the 2nd Guard division. The figure shown here has the full lapels of the 2nd division which they may not have adopted until after the wars. They may have worn their previous line coatee with the addition of the petlitzi (collar and cuff lace) until 1815, as a contemporary Parisian picture shows. The Pavlovski Guard were the only regiment to retain the mitre cap after the 1805 reforms and indeed kept them until 1917 (some with holes received in battles of the Napoleonic wars). Although modern pictures often show chin-scales they were probably not used on mitre caps until after the War, in fact Knötel states this. (a) Shows the back of the mitre cap. Pavlovski officers wore the guard 1812 kiwer or bicorn, not adopting the mitre until May 1825. As mentioned before, the shinel would have the collars (minus petlitzi) shoulder straps and probably cuffs in the regimental colours.

11. Details

The 1809 kiwer (a) had an additional leather flap at the back and dark green cloth ear protectors which folded up inside.

The 1912 kiwer (b) is sometimes shown with a black oilskin or natural linen cover.

The grenadier sergeant (c) has additional gold galloon (tape 5/8in wide) on his collars and cuffs as well as the distinctive top to the plume. All other arms of the army used the same gold galloons to designate NCO's.

The back-view of the infantryman (d) shows the single grenade on the cartouche pouch of the Infantry branch (musketeer).

(e) Shows the triple flamed grenade device of the grenadiers

(f) Shows the jager cartouche box with a regimental number.

12. Shoulder Straps

The number on the straps indicate the division the regiment was in, whereas the colours, in the case of the five infantry (musketeers) straps on the left, indicate seniority in the division, red being the most senior. Jagers had either light blue or yellow and grenadier regiments had red.

13. Pom-poms, Fatigue Caps and Sword Knots

These were the same for infantry and jagers apart from the colour of the fatigue cap headband.

There were three infantry (musketeer) companies to each battalion as shown by the sword knots.

The marksmen formed only a half company (Platoon) in the Infantry and Jager battalions.

14. The Guard 1813-15

The 1st Division comprised of:

1st Brigade (a) Preobrazhenski Lifeguard Regiment (b) Semenovski Lifeguard Regiment

2nd Brigade (c) Ismailovski Lifeguard Regiment (d) Lifeguard Jagerski Regiment Guard Equipage

The 2nd Division comprised of:

1st Brigade (e) Liovski Lifeguard Regiment (f) Lieb-Guard Grenadier Regiment

2nd Brigade (g) Pavlouski Guard Grenadier Regiment (h) Finlandski Lifeguard Jager Regiment Guard Sapper Battalion

Each regiment had three battalions. The Guard Jager Regiments contained Carbinier companies in their battalions (one in each), who wore the grenadier plumes, the rest of the guard Jager companies had pom-poms, as in the Line Jagers. The Guard Jager Regiments coattail turnbacks were green, piped in their distinctive colours; the rest of the Guard had all-red turnbacks. All regiments shoulder straps were red (m) except the Lifeguards Jagerski, which were orange.

Drummer plumes (j) were red, as in the line grenadiers and their lace (k) correspond to the Petlitzi colours of the regiment.

The cartouche pouch (l) show the guard device of the Star of St. Andrew, and four grenades.

All guard buttons and chin-scales were made of copper. Officers had gold pertilzi of various intricate designs according to the regiment. Field officers saddlecloths were the same as the line, except for the Ismailovski and Pavlovski Regiments, which had a white stripe between the double gold stripes (the rest of the Guard had green). The holster caps and lower rear corners of the saddlecloths were decorated by a Silver Star of Andrew.

Russian Infantry deployed in their favoured column of attack formation.

Russian Army (1812-14)

The torment of precautions often exceeds the dangers to be avoided.
It is sometimes better to abandon one's self to destiny.

— Napoleon Bonaparte

The army that faced Napoleon's gigantic invasion force exemplified all the major attributes of the Russian Napoleonic soldier: strength, determination and the ability to endure unbelievable hardship with iron discipline. Their equipment was poor (even by the standards of the period), supplies were problematic, good quality muskets and powder were constantly in demand and food shortages were common. Russian officers were amongst the worst in Europe, often poorly educated and almost illiterate. Despite all this, the Russian army drove the French invaders out of their homeland and kindled the fire of liberation in Western Europe.

Among the Russian infantry much emphasis was put on the power of the bayonet over that of volley fire. The cavalry rode horses whose endurance of local conditions was far superior to that of the invader's mounts. Russian Guards were some of the tallest and bravest soldiers in Europe, specially selected for their size and stamina. The Russian army always had an abundance of cavalry and artillery and contained an almost inexhaustible supply of Cossacks. The Cossacks themselves were some of the most unusual cavalry Napoleon's forces ever faced, fierce warriors who had changed little in appearance and fighting tactics since the Middle Ages.

Russian Army Special Rules

The following rules apply to Russian armies:

Russian Determination

The army of 1812 was famous for its dogged determination. All Russian infantry ignore the first close combat casualty each turn, although the score is counted towards determining the combat winner.

Follow-up casualties after a bout of close combat with other infantry never exceed 1D3 casualties.

Artillery

Russian armies of the period contained huge amounts of artillery. Artillery batteries always contain six gun models (four guns and two howitzers). If you are playing a small battle or have insufficient units for gun allowance, artillery can be chosen at a half battery size of three guns (two guns and one howitzer) at half the normal points cost. Their morale is reduced to 3D6.

Militia Pikemen

Militia pikemen are the most poorly trained equipped units in all the Russian army; they are little more than peasants with stick. Pikemen units do not count towards your PSV score. They fight as normal, but can only form columns or squares and have no firing capacity.

Cossacks

These legendry horsemen harried Napoleon's armies during the great retreat from Moscow. They were free spirited, often raucous and deadly when fighting against stragglers or defeated enemy. They were less effective in a frontal assault on formed troops.

Each infantry division has a free Pulk, or squadron, of Cossacks, represented by three to six models. Cossacks and irregular cavalry were frequently not as controllable as their commanders would have liked so when Cossacks are within engagement range of any enemy troops they are always put on lost command orders.

Some Cossacks can be even more irregular and outdated than normal, like the bow and arrow armed Bashkirs. It is worth having one pulk of them in your army just for variety! The bows have a range of 10" and the unit rolls 1D6 in any direction. A roll of 6 is required to score a hit.

Guard Cossacks do not really count as Cossacks as they are more organised and are lance armed.

Central Russia 1812.

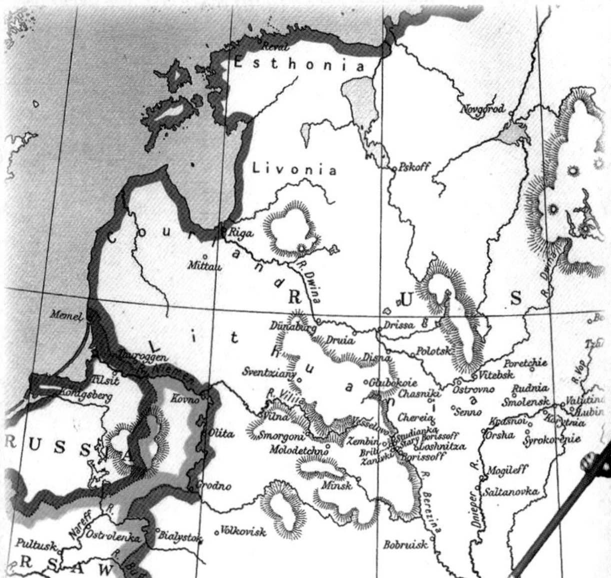

Russian Militia Pikeman.

RUSSIAN ARMY (1812-14)

DIVISIONS, BRIGADES & BATTERIES

Your army may include:

Type	Availability	Consisting of:
Infantry Division	Your army may include any number of infantry divisions.	3-6 infantry units, plus a divisional general; up to a third of the units in each division may be Jaegers or militia. The division may also include one militia pike unit for every militia unit it contains. Each division includes a pulk of Cossacks.
Grenadier Division	Your army may include one grenadier division for every two infantry divisions it contains.	3-4 grenadiers units, plus a divisional commander.
Guard Division	Your army may include a single guard division if it contains at least three infantry divisions.	2-4 guard infantry units, plus a divisional commander.
Guard Cavalry Brigade	Your army may include up to two guard cavalry brigades if it includes a guard infantry division.	1-2 guard cavalry units, plus a brigade commander.
Guard Cavalry Division	Two guard cavalry brigades may be combined to form a guard cavalry division; you must meet the requirements for fielding both brigades.	As brigades; one brigade commander is promoted to divisional general.
Light Cavalry Brigade	Your army may include one light cavalry brigade for each infantry division it contains.	1-2 light cavalry units, plus a brigade commander.
Cuirassier Brigade	Your army may include a single cuirassier brigade if it contains a grenadier division.	1-2 cuirassier units, plus a brigade commander.
Cavalry Division	Your army may include one cavalry division for every two infantry divisions it contains; a cavalry division count as two brigade choices.	Two cavalry brigades of 1-2 units; one brigade may consist of heavy cavalry, the other must consist of light cavalry. Both units in a brigade must be of the same type. One brigade has a brigade commander; the other has a divisional general.

Russian Cuirassiers and Dragoons.

Cossack Brigade	Your army may include a single Cossack brigade if it contains at least three infantry divisions.	1-2 Cossack units, plus a divisional commander.
Irregular Cavalry Units	Your army may include one irregular cavalry unit for every cavalry division contains.	1 irregular cavalry unit.
Foot Artillery Battery	Your army may include one artillery battery for every five infantry units it contains. If your army includes four or less units of infantry, you may include a half battery at half the points cost.	4 foot artillery guns & 2 howitzers. Half battery consist of 2 foot artillery guns & 1 howitzer.
Heavy Artillery Battery	One foot artillery battery me upgraded to heavy artillery for every grenadier or guard division in the army.	This is an upgrade and does not provide the army with any new units: the heavy artillery battery replaces one foot artillery battery and consists of 4 heavy artillery guns & 2 howitzers.
Horse Artillery Battery	Your army may include one horse artillery battery for every four cavalry units in the army (not counting Cossacks or irregular cavalry). If your army contains three or less units of cavalry, you may include a half battery at half the points cost.	Either: 6 horse artillery guns, or 4 horse artillery guns & 1 howitzer. Half batter consists of either: 3 horse artillery guns, or 2 horse artillery guns & 1 howitzer.
Guard Artillery	One of your foot artillery batteries may be upgraded to guard artillery if your army includes a guard infantry division.	This is an upgrade and does not provide any additional units to the army: the upgraded battery becomes a guard artillery battery.
Guard Horse Artillery	One of your horse artillery batteries may be upgraded to guard horse artillery if your army includes a guard cavalry brigade.	This is an upgrade and does not provide any additional units to the army: the upgraded battery becomes a guard horse artillery battery.

Units

Choose the units that make up your divisions from the following:

Troop Type	Points Cost	Unit Size	Morale/Class
Line Infantry	15 points per unit	6	Line Infantry
Jaegers	20 points per unit	6	Light Infantry
Grenadiers	25 points per unit	6	Elite Infantry
Guard Infantry	30 points per unit	6	Guard Infantry
Militia	10 points per unit	6	Militia Infantry
Militia Pikemen	5 points per unit	6	Militia Infantry
Guard Jaegers	35 points per unit	6	Guard Light Infantry
Hussars	35 points per unit	6	Light Cavalry
Lancers	40 points per unit	6	Light Cavalry
Regular Cossacks	40 points per unit	6	Light Cavalry
Irregular Cavalry	20 points per unit	6	Militia Cavalry
Dragoons	40 points per unit	8	Heavy Cavalry
Cuirassiers	45 points per unit	8	Heavy Cavalry
Guard Cossacks	55 points per unit	6	Guard Light Cavalry
Guard Dragoons	55 points per unit	8	Guard Heavy Cavalry
Guard Cuirassiers	60 points per unit	8	Guard Heavy Cavalry
Guard Hussars	50 points per unit	6	Guard Light Cavalry
Mounted Jaegers	35 points per unit	6	Light Cavalry
Foot Artillery Battery	90 points	4 foot & 2 howitzers	Artillery
Heavy Artillery Battery	120 points	4 heavy & 2 howitzers	(Positional) Artillery
Horse Artillery	100 points	6 horse, or 4 horse & 2 howitzers	Artillery
Guard Foot Artillery Battery	+15 points	-	Guard Artillery
Guard Horse Artillery Battery	+15 points	-	Guard Horse Artillery

The Invasion of Russia (1812) French

The French army that invaded Russia was intended to remove the last threat to the Empire in the East. Russia had tried to modernise her armies encouraged by Britain, who was now becoming the Paymaster General of Europe.

To match such a massive undertaking a vast army was required; troops were conscripted from all over Europe and even Austria and Prussia were forced to provide men. Unused as these troops were to the harsh Russian winter, fatigue, the weather and disease caused far more casualties than Russian bullets.

All the French rules as listed in the Peninsular Army Lists apply to this army. Bavarian and Wurtemburg troops (use the Bavarian list) can directly replace any infantry or cavalry divisions.

Divisions, Brigades & Batteries

Your army may include:

Type	Availability	Consisting of:
Infantry Division	Your army may include any number of infantry divisions.	3-5 infantry units, plus a divisional general; up to two units may be light infantry.
Cavalry Brigade	Your army may include one cavalry brigade for each infantry division in the army.	1-2 units of cavalry; both units must be of the same type, except for cuirassiers and carabineer units which may be combined. At least half the cavalry units in the army must be light cavalry.
Cavalry Division	Two cavalry brigades may be combined into a cavalry division. You must meet the requirements for fielding both brigades.	As brigades; both brigades must be of the same type, with the exception of cuirassiers and carabineer units which may be mixed within a division. One brigade commander is promoted to divisional general.
Artillery Battery	Your army may include one artillery battery for every four infantry units it contains. If your army contains three infantry units or less, you may include a half battery at half the points cost.	The first artillery battery in your army must be a foot artillery battery, consisting of 3 foot artillery guns & 1 howitzer. Further artillery batteries may be either foot artillery batteries, or heavy artillery batteries consisting of 3 heavy artillery guns & 1 howitzer.
Horse Artillery Battery	Your army may include one horse artillery battery for every cavalry division it contains, or four every four units of cavalry which are not part of a division.	3 horse artillery guns & 1 howitzer.
Young Guard Division	Your army may include any number of Young Guard divisions if it contains at least two infantry divisions. Infantry divisions must always outnumber the combined total of all guard infantry divisions.	3-4 Young Guard infantry units; up to half the units may be Young Guard light infantry.
Guard Division	Your army may include any number of old guard divisions if it contains at least three infantry divisions. Infantry divisions must always outnumber the combined total of all guard infantry divisions.	2-4 Old Guard infantry units, plus an Old Guard divisional general.
Guard Cavalry Brigade	Your army may include a single guard cavalry brigade if it includes a guard division or at least three infantry divisions. If your army includes two or more guard divisions, you may include two guard cavalry brigades.	The brigade may be either light or heavy. Guard light cavalry brigades consists of 1-3 units of guard light cavalry; guard heavy cavalry brigades consist of 1-2 units of guard heavy cavalry, plus a divisional general (not a brigade commander) in both cases.

Guard Artillery	You may upgrade artillery and horse artillery batteries to guard artillery if you army includes a guard infantry or cavalry division.	This is an upgrade and does not bring your army any additional units: any number of artillery batteries or horse artillery batteries can be upgraded to guard status.

Units

Choose the units that make up your divisions from the following:

Troop Type	Points Cost	Unit Size	Morale/Class
Line Infantry	20 points per unit	6	Line Infantry
Light Infantry	25 points per unit	6	Line Infantry
Young Guard Infantry	25 points per unit	6	Elite Infantry
Young Guard Light Infantry	30 points per unit	6	Elite Light Infantry
Old Guard Infantry	40 points per unit	6	Old Guard Infantry
Light Cavalry	35 points per unit	6	Light Cavalry
Lancers	40 points per unit	6	Light Cavalry
Dragoons	40 points per unit	8	Heavy Cavalry
Cuirassiers	45 points per unit	8	Heavy Cavalry
Carabineers	45 points per unit	8	Heavy Cavalry
Guard Light Cavalry	50 points per unit	6	Guard Light Cavalry
Guard Lancers	55 points per unit	6	Guard Light Cavalry
Guard Heavy Cavalry	55 points per unit	8	Guard Heavy Cavalry
Foot Artillery Battery	60 points/30 points	3 foot artillery & 1 howitzer	Foot Artillery
Heavy Artillery Battery	75 points	3 heavy artillery & 1 howitzer	Heavy Artillery
Horse Artillery Battery	75 points	2 horse artillery & 1 howitzer	Horse Artillery
Guard Foot Artillery	+10 points	-	Guard Foot Artillery
Guard Heavy Artillery	+10 points	-	Guard Heavy Artillery
Guard Horse Artillery	+10 points	-	Guard Horse Artillery

French squares and columns.

French Infantryman.

THE FRENCH ARMY 1812-15

Generals who save troops for the next day are always beaten.

— Napoleon Bonaparte

FRENCH LINE INFANTRY 1812-15

The famous Bardin regulations of 1812 made considerable changes to the appearance of the French infantry. The 'habit long', still in essence the coat of the revolutionary wars, was in theory replaced by the shorter square fronted jacket called the 'habit veste', with a new waistcoat, cut straight across the bottom, replacing the previous version. All this took years to implement, so the old and new uniforms existed concurrently through much of the closing years of empire of the war.

Variations on the new Fusilier dress included: sometimes discs rather than pompoms on the shako, these could have a white centre denoting the number of the battalion, or tufted pompoms (8th) and white or blue piping to the red collar. These troops in Bardin regulations are ideal for many of Napoleon's allies. In 1809 the Saxons adopted French style equipment and appointments, their new type of coat predating Bardin, the same case applies with the Westphalians, Neapolitans and the troops of the Duchy of Cleve-Berg. The shako plates for the Neapolitans and Saxons need to be painted to look like a shield, however the lozenge style plate is correct for Westphalia and Cleve-Berg. The Polish troops of the Vistula legion can also be constructed using these figures if the shako plate is removed and one in the style of a sunburst (similar to those found on Polish czapkas) painted on.

Unit of late French line infantry showing a mix styles.

Unit of late French line infantry.

VOLTIGEURS IN FULL DRESS 1812-15

Having lost their sabre briquet and epaulettes, Voltigeurs were now equipped much as the Fusiliers with only their plumes and moustaches altering the silhouette. They now sported shoulder straps of yellow piped blue (the Fusilier's were blue piped red) and a shorter plain yellow tufted sprang from their unadorned shakos. This, of course, was the official line! In reality they fought hard to hold onto both epaulettes and sabre briquet and the plumes were often green and yellow. Almost universal was the adoption of yellow decorations on the shako including the now ubiquitous yellow Vs on the side.

Voltigeurs in Full Dress.

Drummers

Another major change wrought by the new regulations, in an attempt to curb the caprice of infantry colonels, was the introduction of an 'Imperial livery' for drummers. The new dress consisted of a medium green single breasted coat decorated on the collar, down the front (with five pointed bars of lace) down the sleeves (with seven horizontal chevrons or sometimes loops) and around the cuffs with a new Imperial lace. The lace was yellow edged red with black vertical bars creating yellow square, which alternately held a crowned "N" or an Imperial eagle. All other distinctions of company were supposed to remain unchanged.

Drummers in the new uniforms.

French Grenadier in Full Dress 1812-15

Unlike the Voltigeurs, the Grenadiers held onto all their elite symbolism in the wake of 'Bardin', however they also took up the attraction of decorating their otherwise plain shakos in the way described above.

French Infantry Campaign Dress 1812-15

The pokalem was the 1812 replacement for the bonnet de police, a flat top cap with ear flaps in blue with red piping and with a regiment number or grenade or hunting horn device on the front. They were probably the only headgear supplied to thousands of 'Marie-Louises' called up to defend 'la patrie' in 1814. 'Marie-Louises' was the nickname for the conscripts of 1814, named so after Napoleon's new wife, daughter of the Emperor of Austria and intended to show the conscripts rawness and naïveté!

French Infantry Campaign Dress 1812-15.

French Hussars 1812-15

During the later years of the Napoleonic wars a new cylindrical style shako was issued to the Hussars and to some Chasseur regiments. This new headgear was taller and, as its name implies, did not taper towards the top like earlier styles. In addition it was often covered with coloured cloth which added to the already gaudy appearance of the Hussars. It was used in conjunction with many of the older patterns of shako. Indeed it was probably not until the Waterloo campaign that it became the standard issue. The colours of the shakos worn at Waterloo were as follows: 1st, black; 4th, red; 5th, red (there is some evidence that a pale blue, matching their dolmans was worn earlier in the wars); 6th, red; 7th, pale green. After 1812 Hussar trumpeters were issued Imperial livery.

French Dragoons 1812-15

The Dragoons were also affected by the new Bardin regulations with the introduction of a 'habit veste' similar in style to those of the infantry to replace the old 'habit long'. In addition there were changes to the helmet with the replacement of the plume with company discs. After 1812 Imperial livery was issued to Trumpeters. The new regulations also removed the bearskin cap and replaced it with the helmet adorned with a red plume, the red epaulettes were retained.

Chasseurs in Habit-Veste 1813-15

After 1813, chasseurs adopted the new habit-veste but with facings as before. As an alternative to the colpack for elite companies infantry grenadier-style shakos with red top and bottom bands and side Vs were worn. Some units, however, retained their colpacks despite the regulation.

1813 –
THE WAR OF LIBERATION

Courage is like love: it must have hope for nourishment

— Napoleon Bonaparte

While Napoleon crossed the Niemen Britain, seized her chance to once again unite the countries of Europe against him. Russia joined a new coalition with Britain, Spain and Portugal. With Napoleon's army in retreat the Prussian contingent of Napoleon's army declared its neutrality by the Convention of Tauroggen in December 1812 and then formally joined the coalition in February 1813. The Austrian contingent, which had remained largely inactive during the 1812 campaign, retreated into Bohemia. Austria, although now linked by marriage to Napoleon, joined the coalition, as did Sweden, which was now ruled by Napoleon's former general Marshal Bernadotte, and several of the German states.

Napoleon wasted no time in raising a new army to face the growing threat. He did so by conscripting men from the levees of 1813 and 1814, and by reducing his armies in Spain and across his Empire. Napoleon returned to Germany and intended to defeat his enemies in detail. He was back to his mercurial self. In a whirlwind campaign he took Dresden and then defeated the Prussians under Wittgenstein at Lutzen (2nd May) and Bautzen (20th-21st May). The allies sought an armistice which Napoleon accepted, as it gave him time to train his young army and time for the allies to reorganise. Confident of his own success he resisted offers of peace if France accepted her historic boundaries of the Rhine and the Pyrenees.

On 12th August Austria formally declared war. Napoleon now faced war on three fronts, from the south by the Austrians under Schwarzenburg, from the east by the Prussians now under Blucher, and from the north by the allied army under Bernadotte. The allies gained early successes against Napoleon's subordinates. Bernadotte beat Oudinot at Grossbeeren, while Blucher beat MacDonald at the Katzbach. Napoleon, however, defeated Schwarzenberg at Dresden on 26th-27th August, but again the lack of experience and resources in his army meant that he failed to capitalise on his victory and let the Austrian army escape.

Facing superior odds, fighting on a number of fronts and relying on a young and untrained army Napoleon could only watch as his fortunes slowly drifted away. Further defeats followed and more of his German allies changed sides. On 8th October, Bavaria, his staunchest ally, left him. Napoleon's enemies were now closing in and surrounded his forces at Leipzig. For three days between 16th-19th October the French tried to hold their perimeter lines as Blucher, Schwarzenberg and finally Bernadotte threw their forces against him. The Saxons, his last German allies, defected during the battle and Napoleon had no alternative but to make his escape and retreat to the Rhine with as many men as he could extricate from the battle. 'The Battle of the Nations' as it became known was perhaps the defining moment of Napoleon's career, his Empire was all but lost and his allies had all deserted him. Napoleon gained a last victory at Hanau at the end of October but French control was now back to where it had been twenty years before, based upon her natural boundaries. The war would now become a defensive one as Napoleon's enemies threatened to invade France herself.

Prussian Columns.

THE PRUSSIAN ARMY

Between a battle lost and a battle won, the distance is immense and there stand empires.
— Napoleon Bonaparte

PRUSSIAN INFANTRY

Following the disaster of 1806 the Prussian infantry was completely reorganised, with a new structure for the infantry regiment and new regiments based on the provinces of Prussia. In 1808 a infantry regiment consisted of two musketeer battalions, a fusilier battalion and two companies of grenadiers; the latter were combined with companies from another regiment to form a grenadier battalion.

Unit of Prussian Infantry.

In the new provincial organisation regiments were no longer known by the name of their 'Chef' but were given a provincial name and numbers.

No.1 1st East Prussian	Dull orange collar/cuffs, white shoulder straps.
No.2 1st Pommeranian	White collar/cuffs, white shoulder straps.
No.3 2nd East Prussian	Dull orange collar/cuffs, poppy red shoulder straps.
No.4 3rd East Prussian	Dull orange collar/cuffs, yellow shoulder straps.
No.5 4th East Prussian	Dull orange collar/cuffs, light blue shoulder straps.
No.6 1st West Prussian	Carmine collar/cuffs, white shoulder straps.
No.7 2nd West Prussian	Carmine collar/cuffs, poppy red shoulder straps.
No.8 The Foot Guards Regiment	Poppy red collar/cuffs, white shoulder straps (two bars of white lace on the collars and cuffs).
No.9 The Life Infantry Regiment	Poppy red collar/cuffs, white shoulder straps.
No.10 The Colberg Infantry Regiment	White collar/cuffs, poppy red shoulder straps.
No.11 1st Silesian	Yellow collar/cuffs, white shoulder straps.
No.12 2nd Silesian.	Yellow collar/cuffs, poppy red shoulder straps.

The new Prussian infantry wore a black felt shako, similar in style to the French, worn with a blue Russian style coat, double-breasted with short tails and red turn-backs. The collar and cuffs were in the provincial colour and the shoulder straps distinguished the individual regiments, as shown in the list above.

On campaign the troops wore black oilskin covers on their shakos (troops serving in Russia, in 1812, painted cockades on their covers, this was effectively a white circle and greatcoats in grey were either worn or carried rolled around the body. Grey trousers with short black gaiters were worn until replaced by gaiter-trousers in 1815 (many units were probably not supplied until after the end of hostilities).

FOOT GUARD & THE BRANDENBURG REGIMENT

In 1813 the Foot Guard were removed from the numbering system and a second regiment formed. This meant that numbers 9-12 were renumbered and a new No.12, the Brandenburg infantry regiment, was formed from two reserve battalions and the third battalion of the 1st West Prussians. The new Brandenburg regiment had poppy red on its collar, cuffs and shoulder strap.

Prussian Infantry.

Napoleonic Prussian Reserve Infantry 1813-15

The Kruemper system of training recruits whilst allowing trained men to leave the army had enabled the Prussians to build up a body of experienced soldiers in expectation of the war against France. When the wars of Liberation began the Prussian were able to form battalions quickly by using this reservoir of talent. As well as the regular battalions, which were slightly increased in number, they were thus able to double the number of battalions by using the reserves, and retain a reasonably high quality of personnel in the process. There were 12 Reserve Infantry Regiments formed by the July of 1813 and these were, in March 1815, upgraded to part of the line infantry, numbering as regiments 13-24. Prussian brigades of 1813-1815 were made up equally of one Line Regiment, one Reserve Regiment and one Landwehr (Militia) Regiment, in addition to a Cavalry Regiment and an Artillery Battery.

Unit of Reserve Infantry.

Initially the Reserve infantry was dressed in short grey 'shell jacket' coats with forage caps, very few regiments had proper shakos and coats. However by the autumn of 1813 much of the Reserve Infantry had been re-equipped with British uniforms of various styles destined for the Portuguese Army. These were dark blue with red or green facings or dark blue with white lace and red facings. The British Rifleman uniforms were dark green, faced black. The other 'Shell Jacket' uniforms were overall grey, the caps having Provincial coloured bands. Often the core of a regiment wore the uniforms of the line regiment to which the reserves were attached.

Unit of Reserve Infantry.

Prussian Landwehr Officers.

Prussian Landwehr 1813-1814

Although plans to raise a militia had been put forward as early as 1805, in actuality nothing was done until the early stages of the War of Liberation when East Prussia revolted against the French. The hastily raised militia battalions, known as Landwehr, were raised from all fit men between the ages of 17 and 40, in fact most recruits were under 25. Some were so poorly equipped as to be only armed with pikes!

The militia uniform consisted of a blue (or sometimes black) 'Litewka' style coat with provincial facing colours and a cap often decorated in the same way. The cap or shako, could have an oilskin cover painted with a Prussian (Maltese) cross, usually in white. Leg-wear could consist of white or grey trousers as well as possibly breeches and gaiters; captured French equipment was also used. The provincial colours were:

East Prussia	Poppy red facings, white buttons
Kurmark and Neumark	Poppy red facings, yellow buttons
West Prussia	Black facings, white buttons
Pommerania	White facings, yellow buttons
Silesia	Yellow facings, white buttons.

New areas of recruitment were gained as the war progressed and three other facing colours came into use:

Westphalia	Green facings, white buttons.
The Elbe District	Light blue, yellow buttons.
Rhineland	Madder red, yellow buttons.

Finally in July of 1813 the Militia were organised into regiments (normally of four battalions) giving the Provinces the following strengths:

East Prussia	5 regiments (20 battalions)
West Prussia	3 regiments (11 battalions)
Pomerania	3 regiments (12 battalions)
Neumark and Kurmark	10 regiments (38 battalions)
Silesia	17 regiments (68 battalions).

Prussian Landwehr.

Prussian Landwehr.

Silesian Landwehr.

Prussian Landwehr.

Prussian Reservist Infantryman.

PRUSSIAN CAVALRY

KUIRASSIERS

DRAGOONS

Prussian Reserve Lancers.

RESERVE LANCERS

HUSSARS

GENERALS & AIDES

ARTILLERY

PRUSSIAN ARMY 1813-14

Revolutions are good times for soldiers of talent and courage.

— Napoleon Bonaparte

The Prussian army was rebuilt in 1813. It was formed around the army created during the French occupation when Prussia was only allowed an army of forty-two thousand men. Napoleon forced the reluctant Prussians to join in the invasion of Russia where, under the leadership of General Yorck, they initially fought gallantly for their French masters. As the painful events of 1812 unfolded, Yorck first declared neutrality then changed sides to fight alongside the Emperor's enemies. The following year Prussia rose up and set out on a campaign of repatriation.

One of the most colourful of Prussia's commanders was given the challenge of leading this army of recruits. His name was Blücher and he inspired his men with a renewed sense of national pride.

SPECIAL RULES

The following special rules apply to Prussian armies:

PRUSSIAN BRIGADES

Prussian brigades contained roughly the same number of men as most other countries' divisions. Each brigade's infantry was ideally made up of one third from regular infantry units, one third from reserve infantry units, and one third from Militia units, although some had a much greater proportion of Militia. Brigades also contained a small cavalry detachment of about two squadrons of Lancers or Hussars, along with its own artillery and engineers. Corps (Korps) were made up of a number of infantry brigades and also some cavalry brigades.

SKIRMISHERS

Prussian line infantry formed in line or column of attack and get 2D6 of skirmisher fire per turn. Reserve units roll just1D6.

RESERVISTS

Reservists have the morale of line troops but fight as militia.

MILITIA PIKEMEN

Pikes were provided when the militia, or Landwehr, were first formed in early 1813. After the British provided muskets pikes disappeared completely from the Prussian army; pike armed units could represent the Home Guard, or Landsturm if necessary.

Militia pikemen units do not count towards your PSV score. They fight as normal, but can only form columns or squares and have no firing capacity.

Prussian squares supported by light cavalry confront French light cavalry.

War of Liberation (1813-14) Prussians

Cavalry

At least a third of all cavalry units – including cavalry squadrons – must be reserve cavalry.

Divisions, Brigades & Batteries

Your army may include:

Type	Availability	Consisting of:
Infantry Brigade	Your army may include any number of infantry brigades.	3-6 infantry units, plus a divisional general. Up to one third of the brigade may be line infantry, of which one unit may be grenadiers or fusilier light infantry; one third of the brigade may be reserves and the remaining units, up to a maximum of two thirds of the brigade, will be militia. Each infantry brigade may also include a single cavalry squadron.
Cavalry Brigade	Your army may include one cavalry brigade for each infantry brigade it contains.	1-2 units of Hussars, Reserve Lancers or Line Lancers; both units must be of the same type. If your army includes at least two infantry brigades, one cavalry brigade may consist of 1-2 units of dragoons. If your army includes at least three infantry brigades, one cavalry brigade may consist of 1-2 units of cuirassiers.
Foot Artillery Battery	Your army may include on foot artillery battery for every six infantry units it contains. If your army contains five units or less you may include a half battery at half the points cost.	3 foot artillery guns & 1 howitzer. Half battery consists of 2 foot artillery guns. If your army contains three or more infantry brigades, one battery may be a heavy artillery battery consisting of 3 heavy artillery guns & 1 howitzer.
Horse Artillery Battery	Your army may include one horse artillery battery for every four cavalry units it contains (excluding cavalry squadrons).	3 horse artillery guns & 1 howitzer.
Guards	You may upgrade all the line infantry units in a single division to Guard status, if you army contains three or more infantry brigades.	This is an upgrade and does not provide any additional units to the army; all the line infantry units in the brigade become Guard.

Prussian Infantry, Landwehr and Reservists.

Units

Choose the units that make up your divisions from the following:

Troop Type	Points Cost	Unit Size	Morale/Class
Line Infantry	20 points per unit	6	Line Infantry
Reserve Infantry	15 points per unit	6	Line Infantry (but fight as Militia)
Grenadiers	25 points per unit	6	Elite Infantry
Guards	+20 points per unit	-	Guard Infantry
Militia Infantry	10 points per unit	6	Militia Infantry
Militia Pikemen	5 points per unit	6	Militia Infantry
Fusiliers	25 points per unit	6	Line Infantry
Jaegers	45 points per unit	6	Guard Infantry
Hussars	35 points per unit	6	Light Cavalry
Lancers	40 points per unit	6	Light Cavalry
Reserve Lancers	25 points per unit	6	Militia Cavalry
Reserve Lancer Squadron	10 points per unit	6	Militia (see below)
Dragoons	40 points per unit	8	Heavy Cavalry
Cuirassiers (no armour)	40 points per unit	8	Heavy Cavalry
Foot Artillery	60 points	3 foot, 1 howitzer	Artillery
Heavy Artillery (positional)	75 points	3 foot, 1 howitzer	Artillery
Horse Artillery	75 points	3 horse, 1 howitzer	Artillery

France and Germany 1813.

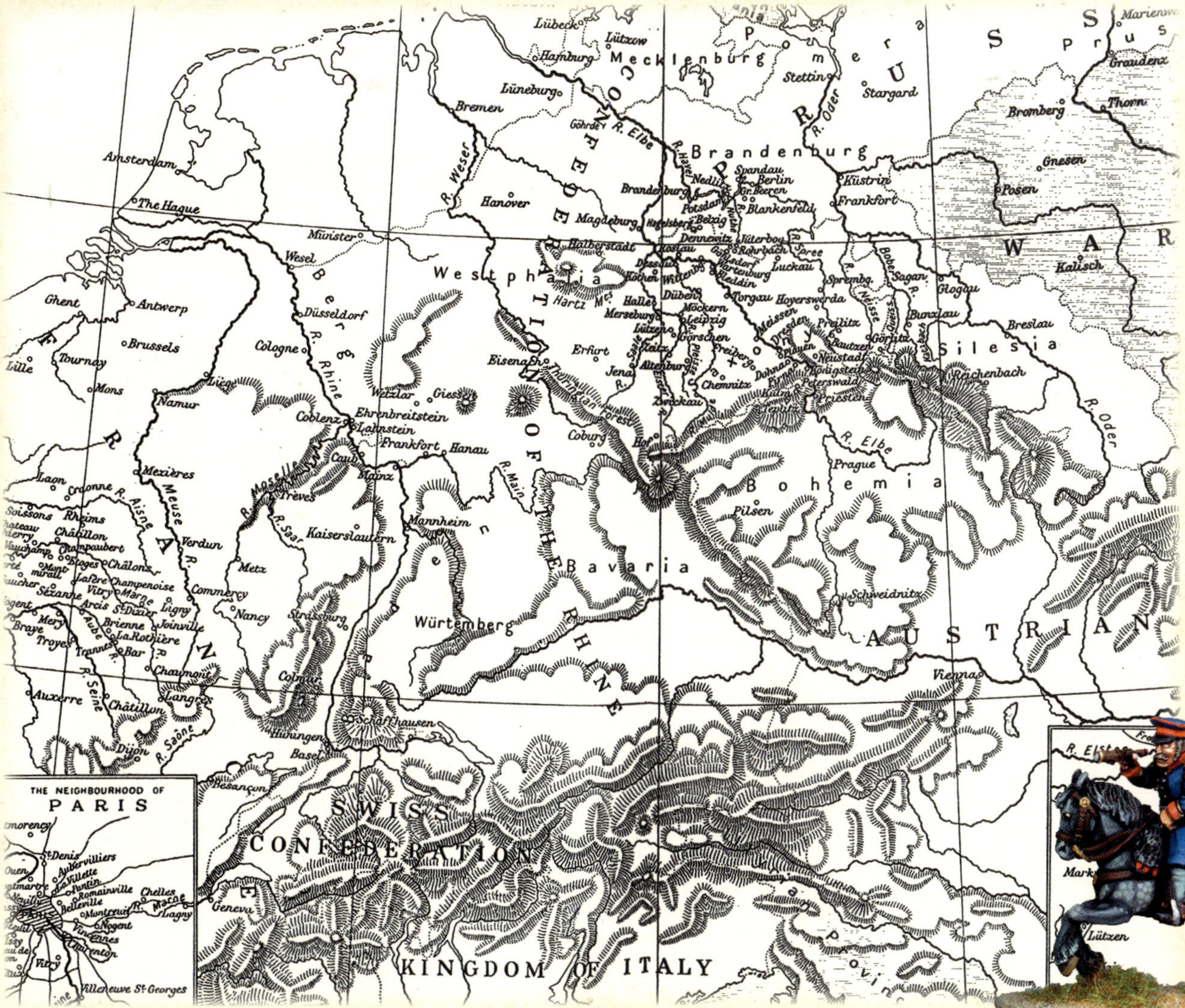

Prussian General.

War of Liberation (1813) French

If the art of war were nothing but the art of avoiding risks, glory would become the prey of mediocre minds. I have made all the calculations, fate will do the rest.

— Napoleon Bonaparte

Special Rules

The normal French rules as described in the Peninsular Wars lists only apply to Guard units in this army.

Line units can still fire their skirmishers but only roll 1D6 for them.

Bavarian or Wurtemburg troops (use the Bavarian list) can directly replace any infantry or cavalry division.

Due to the lack of cavalry the scouting points for this army are added up for the battle as normal and then divided by two.

Divisions, Brigades & Batteries

Your army may include:

Type	Availability	Consisting of:
Infantry Division	Your army may include any number of infantry divisions.	3-5 infantry units, plus a divisional general; up to two units may be light infantry and at least half of the units in each division must be Marie Louise units.
Cavalry Brigade	Your army may include one cavalry brigade for each infantry division in the army.	1-2 units of cavalry; both units must be of the same type, except for cuirassiers and carabineer units which may be combined. At least a third of the cavalry units in the army must be light cavalry or dragoons.
Cavalry Division	Two cavalry brigades may be combined into a cavalry division. You must meet the requirements for fielding both brigades.	As brigades; both brigades must be of the same type, with the exception of cuirassiers and carabineer units which may be mixed within a division. One brigade commander is promoted to divisional general.
Artillery Battery	Your army may include one artillery battery for every four infantry units it contains. If your army contains three infantry units or less, you may include a half battery at half the points cost.	The first artillery battery in your army must be a foot artillery battery, consisting of 3 foot artillery guns & 1 howitzer. Half battery consists of 2 foot artillery guns. Further artillery batteries may be either foot artillery batteries, or heavy artillery batteries consisting of 3 heavy artillery guns & 1 howitzer.
Horse Artillery Battery	Your army may include one horse artillery battery for every cavalry division it contains, or for every four units of cavalry which are not part of a division.	3 horse artillery guns & 1 howitzer.
Young Guard Division	Your army may include any number of Young Guard divisions. infantry divisions. Infantry divisions must always outnumber the combined total of all guard infantry divisions.	3-4 Young Guard infantry units, plus a divisional general; up to half the units may be Young Guard light infantry.
Guard Division	Your army may include any number of old guard divisions if it contains at least three infantry divisions. Infantry divisions must always outnumber the combined total of all guard infantry divisions.	2-4 Old Guard infantry units, plus an Old Guard divisional general.

Guard Cavalry Brigade	Your army may include a single guard cavalry brigade if it includes a guard division or at least four infantry divisions. If your army includes two or more guard divisions, you may include two guard cavalry brigades.	The brigade may be either light or heavy. Guard light cavalry brigades consists of 1-3 units of guard light cavalry; guard heavy cavalry brigades consist of 1-2 units of guard heavy cavalry, plus an Old Guard divisional general (not a brigade commander) in both cases.
Guard Artillery	You may upgrade artillery and horse artillery batteries to guard artillery if you army includes a guard infantry or cavalry division.	This is an upgrade and does not bring your army any additional units: any number of artillery batteries or horse artillery batteries can be upgraded to guard status.

UNITS

Choose the units that make up your divisions from the following:

Troop Type	Points Cost	Unit Size	Morale/Class
Marie Louise	10 points per unit	6	Militia Infantry
Line Infantry	15 points per unit	6	Line Infantry
Light Infantry	20 points per unit	6	Line Infantry
Young Guard Infantry	20 points per unit	6	Elite Infantry
Young Guard Light Infantry	35 points per unit	6	Elite Light Infantry
Old Guard Infantry	40 points per unit	6	Old Guard Infantry
Light Cavalry	35 points per unit	6	Light Cavalry
Lancers	40 points per unit	6	Light Cavalry
Dragoons	40 points per unit	8	Heavy Cavalry
Cuirassiers	45 points per unit	8	Heavy Cavalry
Carabineers	45 points per unit	8	Heavy Cavalry
Guard Light Cavalry	50 points per unit	6	Guard Light Cavalry
Guard Lancers	55 points per unit	6	Guard Light Cavalry
Guard Heavy Cavalry	55 points per unit	8	Guard Heavy Cavalry
Guards of Honour	45 points per unit	6	Guard Light Cavalry
Foot Artillery Battery	60 points/30 points	3 foot artillery & 1 howitzer	Foot Artillery
Heavy Artillery Battery	90 points	3 heavy artillery & 1 howitzer	Heavy Artillery
Horse Artillery Battery	75 points	2 horse artillery & 1 howitzer	Horse Artillery
Guard Foot Artillery	+10 points	-	Guard Foot Artillery
Guard Heavy Artillery	+10 points	-	Guard Heavy Artillery
Guard Horse Artillery	+10 points	-	Guard Horse Artillery

French Guard Horse Artillery and Horse Grenadiers.

1814 – The War for France

In war, the general alone can judge of certain arrangements. It depends on him alone to conquer difficulties by his own superior talents and resolution.

France was facing invasion on all fronts with the Prussians, Russians and Austrians threatening her via Germany, Switzerland and Italy, while the British, Spanish and Portuguese threatened her Spanish border. These Allies again offered Napoleon the chance of peace if he would accept the traditional, physical boundaries of France. He again refused – pride or belief in his own invincibility drove him on to defend France.

Napoleon had raised a large army but it was no longer *le grande armée* of 1805, composed of well-trained and seasoned soldiers. Instead his army consisted of many youngsters conscripted before their time, (known as the 'Marie-Louises' after Napoleon's young wife), and invalids drawn from the garrisons. His commanders were also growing old, consisting mainly of those that still remained from the devastating wars of the previous years. But if his army was tiring Napoleon was not, and he threw himself into the conflict with renewed vigour. The greater the challenge the higher he appeared to rise to meet it.

French Infantry (Knötel).

French Guard (Knötel).

His strategy was again to prevent the unification of the Allied armies by defeating each in turn. The Allies had anticipated his plans though, and had agreed that each would fall back in the face of Napoleon, and advance in his absence. The war started early in 1814. Having left instructions for Paris to be fortified and defended, Napoleon commenced his campaign against Blucher's Prussians gaining victory after victory in quick succession: at Brienne on 29th January, at La Rothiere the next day and at Champaubert on 10th February. This was followed by Montmirail on the 11th February, Chateau-Thierry on the 12th and Vauchamps on the 14th. With the Prussians in retreat he could now turn to Schwarzenberg's Austrians.

Again he was victorious, this time at Montereau on the 18th February, but on hearing that Blucher had resumed his advance and was within twenty five miles of Paris, Napoleon left MacDonald to watch Schwarzenberg while he switched his efforts back against the Prussians; he defeated them at Craonne on 7th March. Napoleon's subordinates were never as great as their master and on 27th February Schwarzenberg defeated MacDonald at Bar-Sur-Aube.

By 1813 the Allies were beginning to unite and Blucher's force was reinforced by Bernadotte thereby giving them a significant numerical advantage which they were able to exploit in their victory over Napoleon at Laon during the 9th and 10th March. Napoleon struck back defeating one of Blucher's corps at Rheims on 13th March before again assailing Scwarzenberg, to whom he lost at Arcis-sur-Aube on the 20th-21st March.

The campaign was now going against Napoleon. The Allies were pushing towards Paris and were growing in numbers while his own forces were growing weaker by marches and counter-marches and by the numbers of dead and wounded which could not be replaced.

Mortier and Marmont sought to halt Schwarzenberg's advance but were defeated at La-Fere-Champenoise, leaving the Austrian General free to link up with Blucher. Napoleon sought to draw his assailants away from Paris but his orders were intercepted and the Allies ignored the ruse and headed straight for Paris. Marmont, ordered to defend the city at all costs, surrendered after a token gesture and on 31st March the capital fell.

Abdication & Exile

Napoleon had been betrayed by his government but intended to fight on. He set up his headquarters at Fontainebleau where he consulted with his remaining Marshals as to what to do next. They persuaded him that continued fighting was pointless and that abdication was the only appropriate course of action. Napoleon attempted to take his life by poison but failed, and on 6th April offered his abdication in favour of his son.

The Allies would not consent to these terms and on 11th April they accepted his unconditional surrender, placing the Bourbons, who had lost the throne in 1793, back into power in the guise of Louis XVIII. Napoleon was exiled to the small island of Elba off the coast of Italy where he was allowed to keep a small army and was to be paid a pension by France. France itself was restored to the boundaries of 1792.

French Swiss Regiment.

Boney and his high command.

THE BAVARIAN ARMY

In war, the moral element and public opinion are half the battle.

— Napoleon Bonaparte

Bavaria fought alongside France from 1805 until 1813, regarding themselves as partners to the French, rather than subservient auxiliaries. The Bavarian army, however, was rather small, with just eleven regiments of infantry, six of light infantry and six of Chevauxlegers.

The armed might of Maximillian Joseph of Bavaria continued the history of Bavarian support for France until the events of 1813 forced a change of alliances. Unfortunately for the Bavarians their first battle on the side of the allies was to be Napoleon's next victory for as they attempted to block the French withdrawal at Hanau they were swept aside. Nevertheless they were generally good soldiers and while they had only taken a secondary role in the campaigns of 1805-7, they were fully committed to both the 1809 campaign and the Russian adventure of 1812.

Unit of Bavarian Line Infantry.

BAVARIAN INFANTRY

Uniforms of the Line and Light Infantry were of the same cut but of distinct colours. The line troops wore bright light blue with facings of poppy red, yellow, pink and black.

The light infantry wore green with facings of poppy red, black, and crimson (changed to yellow in 1811).

In 1814 all facings became red. The difference in officers' dress from 1812 was the adoption of the gorget in preference to the traditional sash of silver worked with blue.

Bavarian Light Infantry.

Bavarian Infantry.

Bavarian army deployed in mixed order formation.

Bavarian Elite Line And Light Infantry

Bavarian Grenadiers wore plumes of red from the introduction of the new uniform in 1799, however the Schutzen, or light, companies did not adopt plumes until 1811, when green ones were issued. For the 2nd battalions the bottom halves of these plumes were white and in the 3rd battalions (from 1814) black.

Bavarian Elite Grenadiers.

Bavarian Chevauxlegers

The history of the Bavarian cavalry is particularly complicated, but it breaks down to the following: in 1804 there were two regiments of Dragoons and four regiments of Chevauxleger. Dragoons dressed in white with red facings while Chevauxleger dressed in light green coats with the following faces:

Regiment	Facings & Details
1st & 4th	Black piped red.
2nd	Poppy red.
3rd	Black.

Bavarian Chevauxlegers.

The first change occurred in 1809 when the colour of the Chevauxleger's uniform became dark green. Then in 1811 the two Dragoon regiments became Chevauxleger, making six in total (the new precedence caused the complete renumbering of the regiments). The two new Chevauxleger regiments retained their red facings.

The uniforms of the Dragoons were identical to those of the Chevauxleger with only the saddlery being different; Bavarian Dragoons had sheepskin saddles. The shabraques for both types of Bavarian were red with a white border decorated with a continuous diamond pattern in blue. Mounted officers shabraque was in the facing colour up till 1810 and red thereafter.

Artillery Pieces

Woodwork for Bavarian Artillery was a bright medium blue.

Bavarian Artillery.

Bavarian Infantry.

Campaign of 1814 – Allied Armies

The art of war is to gain time when your strength is inferior.

— Napoleon Bonaparte

The allied armies of 1814 were composite forces derived from the many nations of Europe united by the shared ambition of finally crushing Napoleon. These armies were much smaller than the colossal forces of previous years. By this time battles were often fought with only a single corps's strength per side. What the allies did not expect at this stage was a reinvigorated Napoleon still capable of giving battle hence, where possible, the allies concentrated their fight against Napoleon's subordinate generals' armies. Eventually this tactic proved successful, finally enabling the allies to march to the heart of the French Empire, Paris.

This campaign allows players to field interesting and varied small armies drawn from multiple nations.

Army of Silesia 1814

Commanded by the Prussian Field Marshal Blücher this army was a combination of Prussians and Russians; so you could collect and deploy a splendid multinational force. Your first division should be made up entirely of Russian or Prussian infantry whilst the second division can consist wholly of troops from either nation – as can subsequent divisions.

Both divisions can field half batteries.

Your army can include Guard or cavalry from the nation of your first division.

So, for example, you started with a Russian infantry division, then fielded two Prussian divisions, you can deploy a Russian Guard division.

Army of Bohemia 1814

This is again a multinational army, led by the Austrian Field Marshal Schwarzenberg but based primarily around the Austrian army. It contained many troops from Napoleon's former German allies such as Bavaria and Wurtemburg.

Your first division must be an Austrian formation.

Subsequent divisions can be the same or consist of German allies (use the Bavarian list).

Rules for elite divisions are the same as for the Army of Silesia.

Bavarian Army 1809-1814

Bavaria and the Confederation of the Rhine c.1810.

The Bavarian army is one of the most colourful of all Napoleonic armies to deploy on the battlefield, resplendent in their famous cornflower blue coats. After being one of France's most dependable allies, Bavaria changed allegiance and joined her enemies; your Bavarian army could fight for either side.

The army had been arranged in the French manner; where infantry were organised along the lines of four Fusilier companies, one Grenadier company and one light company. The whole army only ever reached corps strength so you could feasibly field every unit in the Bavarian army!

With a few minor alterations this Army List can be used to represent any other of the German states, as long as all players agree with the changes made.

DIVISIONS, BRIGADES & BATTERIES

Your army may include:

Type	Availability	Consisting of:
Infantry Divisions	Your army may include any number of infantry divisions.	3-5 infantry units, plus a divisional general; up to two units may be light infantry.
Cavalry Brigade	Your army may include one cavalry brigade for each infantry division it contains.	1-2 light cavalry units, plus a brigade commander.
Cavalry Division	Two cavalry brigades may be combined to form a cavalry division; you must meet the requirements to field both brigades.	As brigades; one brigade commander is promoted to divisional general.
Artillery Battery	Your army may include one foot or heavy artillery battery for every four infantry units it contains. If your army includes three infantry units or less you may include a half battery at half the points cost.	Foot artillery batteries consist of 3 foot artillery guns & 1 howitzer. Half battery consists of 2 foot artillery guns. Heavy artillery batteries consist of 3 heavy artillery guns & 1 howitzer. Half battery consists of 2 heavy artillery guns.
Horse Artillery Battery	Your army may include one horse artillery battery for every cavalry division it contains.	2 horse artillery guns & 1 howitzer.

The composition of the Bavarian army changed only slightly during the course of the Napoleonic wars:

1809 Campaign

- Up to half your infantry units can be of elite status. No conscript units are allowed.
- One of your cavalry brigades may be made up of 1-2 units of dragoons.

1813 Campaign

- Up to half your infantry can be conscript units.
- One of your cavalry brigades may be made up of 1-2 units of dragoons.

1814 Campaign

- You can include a maximum of one Guard unit if your army has at least two infantry divisions.
- Up to half your infantry can be conscript units
- One of your cavalry brigades may be made up of 1-2 units of dragoons. One of these dragoon units may be replaced by a cuirassier unit.

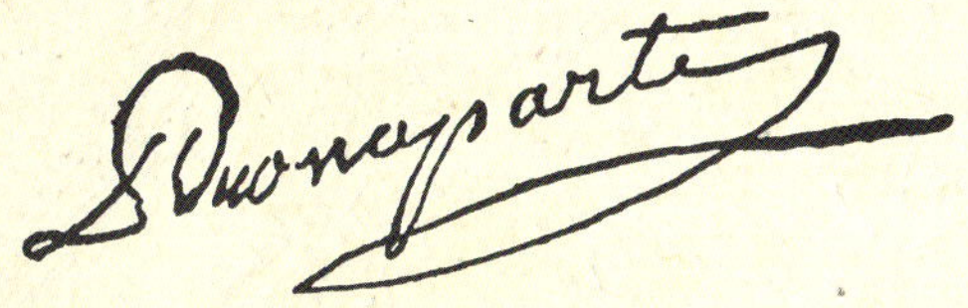

UNITS

Choose the units that make up your divisions from the following:

Troop Type	Points Cost	Unit Size	Morale/Class
Conscripts	15 points per unit	6	Militia
Line Infantry	20 points per unit	6	Line
Light Infantry	25 points per unit	6	Line
Elite Line Infantry	25 points per unit	6	Elite
Elite Light Infantry	30 points per unit	6	Elite
Guard Grenadiers	40 points per unit	6	Guard
Light Cavalry	35 points per unit	6	Light Cavalry
Dragoons	40 points per unit	8	Heavy Cavalry
Cuirassiers	45 points per unit	8	Heavy Cavalry
Foot Artillery Battery	60 points	3 foot, 1 howitzer	Artillery
Horse Artillery Battery	75 points	2 horse, 1 howitzer	Artillery
Heavy Artillery Battery	90 points	3 heavy, 1 howitzer	Artillery

Following the defeat of Leipzig, Napoleon faced a renewed threat and, for the second time in two years, needed to build a new army. Not since the Revolutionary Wars had French armies needed to fight in defence of France herself.

It would be a joke if the conduct of the victor had to be justified to the vanquished.

— Napoleon Bonaparte

French Grenadier a Cheval (Knötel).

French Grenadier Officer.

Campaign for France 1814

War is like government, a matter of tact.

— Napoleon Bonaparte

The new army suffered far worse shortages of everything it needed to function than its predecessor. Even the most basic equipment was becoming scarce. France was increasingly war weary and despondent after two years of defeat. The Napoleonic dream was fading.

Faced with adversity Napoleon's confidence returned and he regained much of his old genius. After some early setbacks at La Rothiere, victory soon followed fast upon victory. Sub commanders like Marshal Ney were now commanding at divisional level; Ney himself commanded the Young Guard. These new officers, directly under the Emperor's control, completely surprised the allies with their military and tactical ability, at a time when the allies had adopted a new policy of targeting French armies commanded by subordinates rather than Napoleon himself!

The National Guard was called to fight alongside the new army, and even more Marie Louise battalions were created to fight next to Old Guard Chasseurs. New squadrons of Guard cavalry, the Imperial Scouts, were formed to bulk out the ever decreasing numbers of cavalry available.

French Army Special Rules

The normal French rules as given in the Peninsular Wars lists only apply to Guard units in this army. Line units can still fire their skirmishers but only roll 1D6 for them.

The lack of cavalry means that scouting points for this army are added up for the battle as normal, and then divided by three.

French Army with a Grand Battery face the Austrians.

WAR FOR FRANCE (1814) FRENCH

Great battles are won with artillery.

— Napoleon Bonaparte

DIVISIONS, BRIGADES & BATTERIES

Your army may include:

Type	Availability	Consisting of:
Infantry Division	Your army may include any number of infantry divisions.	3-5 infantry units, plus a divisional general; up to two units may be light infantry and at least three-quarter of the units in each division must be Marie Louise or National Guard units.
Cavalry Brigade	Your army may include one cavalry brigade for every two infantry divisions in the army.	1-2 units of cavalry; both units must be of the same type, except for cuirassiers and carabineer units which may be combined. At least a third of the cavalry units in the army must be light cavalry or dragoons.
Cavalry Division	Two cavalry brigades may be combined into a cavalry division. You must meet the requirements for fielding both brigades.	As brigades; both brigades must be of the same type, with the exception of cuirassiers and carabineer units which may be mixed within a division. One brigade commander is promoted to divisional general.
Artillery Battery	Your army may include one artillery battery for every six infantry units it contains. If your army contains five infantry units or less, you may include a half battery at half the points cost.	The first artillery battery in your army must be a foot artillery battery, consisting of 3 foot artillery guns & 1 howitzer. Half battery consists of 2 foot artillery guns. Further artillery batteries may be either foot artillery batteries, or heavy artillery batteries consisting of 3 heavy artillery guns & 1 howitzer.
Horse Artillery Battery	Your army may include one horse artillery battery for every cavalry division it contains, or for every four units of cavalry which are not part of a division.	3 horse artillery guns & 1 howitzer.
Young Guard Division	Your army may include any number of Young Guard divisions. Infantry divisions must always outnumber the combined total of all guard infantry divisions.	3-4 Young Guard infantry units, plus a divisional general; up to half the units may be Young Guard light infantry.
Guard Division	Your army may include a single Old Guard divisions if it contains at least three infantry divisions. Infantry divisions must always outnumber the combined total of all guard infantry divisions.	2-4 Old Guard infantry units, plus an Old Guard divisional general.
Guard Cavalry Brigade	Your army may include a single guard cavalry brigade if it includes a guard division or at least four infantry divisions. If your army includes two or more guard divisions, you may include two guard cavalry brigades.	The brigade may be either light or heavy. Guard light cavalry brigades consists of 1-3 units of guard light cavalry; guard heavy cavalry brigades consist of 1-2 units of guard heavy cavalry, plus a brigade commander in both cases. Any Guard cavalry brigade consisting of two or three units can be a Guard of Honour, in which case the brigade commander is promoted to Old Guard divisional general.
Guard Artillery	You may upgrade artillery and horse artillery batteries to guard artillery if you army includes a guard infantry or cavalry division.	This is an upgrade and does not bring your army any additional units: any number of artillery batteries or horse artillery batteries can be upgraded to guard status.

UNITS

Choose the units that make up your divisions from the following:

Troop Type	Points Cost	Unit Size	Morale/Class
Pupils/Velites	25 points per unit	6	Elite
National Guard	5 points per unit	6	Militias
Marie Louise	10 points per unit	6	Militia Infantry
Line Infantry	15 points per unit	6	Line Infantry
Light Infantry	20 points per unit	6	Line Infantry
Young Guard Infantry	25 points per unit	6	Elite Infantry
Young Guard Light Infantry	30 points per unit	6	Elite Light Infantry
Old Guard Infantry	40 points per unit	6	Old Guard Infantry
Light Cavalry	35 points per unit	6	Light Cavalry
Line Lancers	40 points per unit	6	Light Cavalry
Dragoons	40 points per unit	8	Heavy Cavalry
Cuirassiers	45 points per unit	8	Heavy Cavalry
Carabineers	45 points per unit	8	Heavy Cavalry
Guard Light Cavalry	50 points per unit	6	Guard Light Cavalry
Guard Lancers	55 points per unit	6	Guard Light Cavalry
Guard Heavy Cavalry	55 points per unit	8	Guard Heavy Cavalry
Guards of Honour	50 points per unit	6	Guard Light Cavalry
Imperial Scouts	40 points per unit	6	Light Cavalry
Foot Artillery Battery	60 points/30 points	3 foot artillery & 1 howitzer	Foot Artillery
Heavy Artillery Battery	90 points	3 heavy artillery & 1 howitzer	Heavy Artillery
Horse Artillery Battery	75 points	2 horse artillery & 1 howitzer	Horse Artillery
Guard Foot Artillery	+10 points	-	Guard Foot Artillery
Guard Heavy Artillery	+10 points	-	Guard Heavy Artillery
Guard Horse Artillery	+10 points	-	Guard Horse Artillery

Napoleon at the Battle of Austerlitz, December 2, 1805.

French Old Guard Drummer.

Command Cards

Napoleon has humbugged me, by God; he has gained twenty-four hours' march on me.
— Duke of Wellington

Command Cards

Napoleon

Charge!

The player orders his unit to charge an enemy unit and engage them in close combat. He nominates the target of the charge then measures the distance between the two units.

Napoleon

Run!

The unit flees its perilous position by running with all possible speed towards its own table edge. The unit turns 180º and be moves three times its normal move distance away from the enemy.

Napoleon

Retreat and Fire!

The unit retreats half a standard move away from the enemy while still facing them, then halts and fires off a volley.

Napoleon

Stand and Fire!

The unit remains stationary and fires a volley at the enemy.
If the unit is being charged the firing is assumed to take place at point blank range.

Napoleon

Retreat!

The unit retreats backwards, still facing the enemy, up to the following distances.

Napoleon

Advance and Fire!

The unit moves its normal move distance towards the enemy, then halts and looses a volley of fire. The advancing unit may change direction during its move.

Napoleon

Form Line!

A unit can be ordered to change formation to line. The unit given a formation order must remain stationary to perform the change of formation and cannot fire.

Napoleon

Form Square!

A unit can be ordered to change formation to square. The unit given a formation order must remain stationary to perform the change of formation and cannot fire.

Napoleon

Form Column!

A unit can be ordered to change formation to square. The unit given a formation order must remain stationary to perform the change of formation and cannot fire.

Napoleon

Unlimbered Artillery Stand and Fire!

The unit remains stationary and fires a volley at the enemy.
If the unit is being charged the firing is assumed to take place at point blank range.

Napoleon

Unlimbered Artillery Run!

The unit flees its perilous position by running with all possible speed towards its own table edge. The unit turns 180º and be moves three times its normal move distance away from the enemy.

Napoleon

Unlimbered Artillery Retreat!

The unit retreats backwards, still facing the enemy, up to the following distances.

Napoleon

Limbered Artillery Retreat!

The unit retreats backwards, still facing the enemy, up to the following distances:

Napoleon

Limbered Artillery Run!

The unit flees its perilous position by running with all possible speed towards its own table edge. The unit turns 180º and be moves three times its normal move distance away from the enemy.

Napoleon

First Volley!

Units making their first volley of the battle can march and fire, but cannot re-roll any misses.

Napoleon

Lost Command!

Lost command units roll 1D6 on the appropriate chart and react as indicated

1815 – The Return

What my enemies call a general peace is my destruction. What I call peace is merely the disarmament of my enemies. Am I not more moderate than they?

— Napoleon Bonaparte

Growing unrest in France, and the failure of Louis XVIII to keep up Napoleon's pension, opened the way to his return. Waiting until the British ship that was to keep watch upon him had returned to port, he slipped away in boats with his small force and landed near Antibes on the south coast of France. In a spectacular march north his army grew as men returned to his colours and the regiments sent to halt him deserted to his cause. He returned to Paris in triumph as Louis XVIII fled.

Overtures for peace were snubbed by the Allied powers, who branded Napoleon an outlaw and immediately made plans to resume hostilities against him and against France. The British and Prussians were the first in the field stationing forces in Belgium, with the Dutch-Belgian forces under Wellington. But unfortunately, Napoleon realised that the Austrians and Russians would take longer to assemble their armies and would arrive at the French border much later.

Seizing the opportunity Napoleon moved his army to the Belgian border, anticipating that if he could gain decisive victories over the Prussians and British he would divide the Allied Powers. He sensed that disagreements between them would result in Austria and Russia agreeing to peace.

Quatre Bras

On the evening of the 14th June his army crossed into Belgium near Charleroi and his main force struck at the Prussians under Blucher at Ligny, while a smaller force held the British at Quatre Bras. Napoleon gained the victory he wanted over the Prussians, although he was unable to bring the decisive numbers to bear and the Prussians retreated. But the pursuit was delayed thereby enabling the Prussians – rather than retreating east away from Wellington's forces – to move towards them.

The Battle of Waterloo, 18th June 1815.

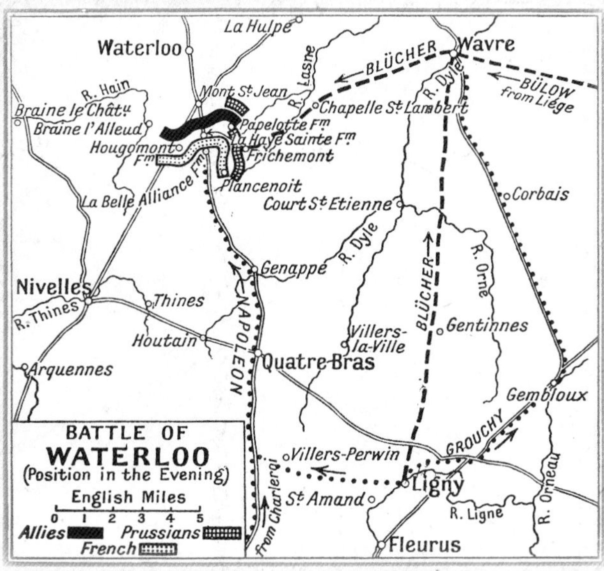

Waterloo

The 17th June was used by Napoleon to re-gather his forces before attacking the British near the village of Waterloo on the 18th June. The wet weather delayed the start of the battle, which descended into a frontal attack on the British and Dutch lines. Wellington, using the tactics of the reverse slope and combating the French assaults by the use of line and square, held his position. During the afternoon the Prussians started to arrive on the edge of the battlefield and Napoleon was faced with having to split his forces to repel the new threat. The arrival of the Prussians was timely as the British line was close to breaking; even a final onslaught by the Imperial Guard could not dislodge it and with the French army now broken, Napoleon fled the field. He still hoped to rally his troops, but finding no support in the capital to sustain the war he had no choice but to abdicate once again, this time to be exiled on the island of St Helena in the South Atlantic where he was to see out his final years.

With Napoleon's defeat at Waterloo, over twenty years of almost continuous war came to an end. The map of Europe was changed, the balance of power had shifted and new ideas of nationalism had grown leading to the eventual unification of Italy and Germany. Warfare was never to be the same again but the legend of Napoleon and his victorious armies would live on for all time.

QUICK REFERENCE SHEET

TURN SEQUENCE

The sequence of phases is as follows:

1. **Roll for Initiative.**
2. **Place Command Cards**
3. **Artillery Fire**
4. **Reveal Command Cards**
5. **Movement**
6. **Skirmisher Fire Resolution**
7. **Volley Fire Resolution**
8. **Close Combat Resolution**
9. **Morale Checks**
10. **Determining the Winner**

ARC OF FIRE

A unit can fire inside an area measured by drawing a line, 45° degrees to either side of its front bases.

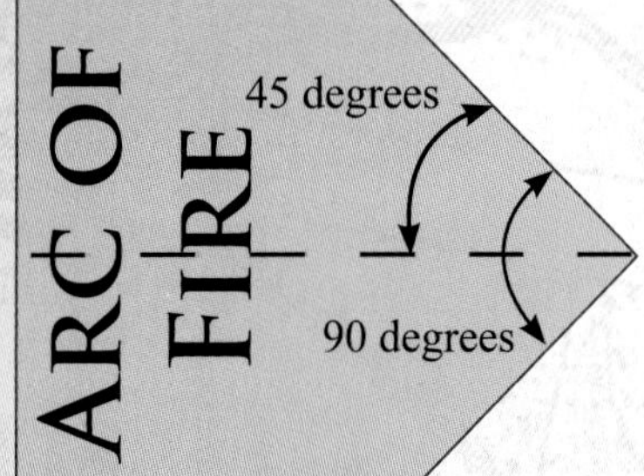

Arc of fire for artillery firing canister is the area within two lines drawn at a 90 degree angle from the centre of **each individual artillery base** (see diagram page 42).

ARTILLERY RANGE

	Close Canister	Canister	Effective Range	Long Range	Extreme Range
Heavy Artillery	0 to 6”	6+ to 12”	0 to 24”	24+ to 36”	36+ to 60”
Field Artillery	0 to 6”	6+ to 12”	0 to 24”	24+ to 36”	36+ to 48”
Horse Artillery	0 to 6”	6+ to 12”	0 to 24”	24+ to 36”	36+ to 48”

Howitzers have the same ranges as field artillery (see Howitzers, page 44).

ENGAGEMENT RANGE

Unit	Standard Movement Rate	Engagement Range
Light cavalry	8”	16”
Heavy cavalry	6”	12”
Infantry	4”	8”
Artillery	4”	8”

ARTILLERY FIRING CANISTER (CASE/GRAPESHOT)

	Close Canister	Canister
Heavy Artillery	3+	4+
Field Artillery	4+	5+
Horse Artillery	4+	5+

ARTILLERY FIRING INTO THE FRONT OR REAR

	Score Needed to Hit		
Target Unit	Effec-tive Range	Long Range	Ex-treme Range
Artillery Battery	6	6*	-
Rifle Detach-ment	6	6	-
Cavalry Line	6	6	6*
Cavalry Column	5+	6	6
Infantry Column	5+	6	6
Infantry Square	4+	5+	6
Infantry Line	4+	5+	6*

ARTILLERY FIRING INTO THE FLANK

	Score Needed to Hit		
Target Unit	Effec-tive Range	Long Range	Ex-treme Range
Artillery Battery	6	6*	-
Rifle Detach-ment	6	6*	-
Cavalry Line	6	6*	-
Cavalry Column	5+	6	6*
Infantry Column	5+	6	6*
Infantry Square	4+	5+	6*
Infantry Line	6	6*	-

MOVEMENT RATES FOR DIFFERENT TROOP TYPES

Unit	Standard	March	Forced March	Charge	Retreat and Fire	Retreat	Advance and Fire	Running
Light Cavalry	8”	16”	24”	16”	4”	8”	8”	24”
Heavy Cavalry	6”	12"	18"	12"	3”	6”	6”	18"
Infantry	4”	8”	12”	8”	2”	4”	4”	12”
Artillery	4”	8”	12”	N/A	N/A	4”	N/A	12”
Horse Artillery	8”	16”	24”	N/A	4”	2”	N/A	24”
Skirmishers	8”	16”	24”	N/A	N/A	8”	N/A	24”

Commanders can always move up to 24” in each Movement phase. They are not affected by any terrain or weather restrictions and cannot become fatigue casualties.

Quick Reference Sheet

Musket Armed Light Companies

Firing Range	Distance	D6 Score Required to Hit
Effective	0-10"	5+
Long	10-20"	6

Rifle Armed Light Companies

Firing Range	Distance	D6 Score Required to Hit
Effective	0-12"	5+
Long	12-24"	6

Firing Formation

Firing Unit's Formation	Number of D6 Unit Can Roll
Line	6 dice
Column	2 dice
Square	2 dice per side of square

Guard units receive one extra firing die per unit.

Volley Fire

	Range and Score Required to Hit			
Weapon	0-5"	5-10"	10-15"	15-20"
Musket	3+	5+	6	-
Carbine	5+	6	-	-

Morale Tests

Troop Type	Number of D6 to Test Morale
Militia	2D6
Line Infantry	3D6 and discard lowest dice roll
Elite Infantry	3D6
Guard Infantry	4D6 and discard lowest dice roll
Militia Cavalry	2D6
Light Cavalry	3D6 and discard lowest dice roll
Elite Light Cavalry	3D6
Heavy Cavalry	3D6 and discard lowest dice roll
Elite Heavy Cavalry	3D6
Guard Cavalry	4D6 and discard lowest dice roll
Artillery	3D6
Guard Artillery	4D6

Lost Command Orders

Militia Infantry Lost Command Orders

D6 Score	Reaction
1	Run.
2-3	Retreat.
4-5	Remain stationary.
6	Unit receives a stand and fire order.

Infantry Lost Command Orders

D6 Score	Reaction
1	Retreat.
2-3	Remain stationary.
4-5	Stand and fire.
6	Advance and fire.

Elite and Guard Infantry Lost Command Orders

D6 Score	Reaction
1	Retreat.
2-3	Retreat and fire.
4-5	Stand and fire.
6	The unit charges.

Cavalry

D6 Score	Reaction
1	Run.
2-3	Retreat.
4-5	Remain stationary.
6	The unit charges.

Elite and Guard Cavalry

D6 Score	Reaction
1	Retreat.
2-3	Remain stationary.
4-6	The unit charges.

Artillery

D6 Score	Reaction
1	Abandon guns and run.
2-3	Limber guns and retreat.
4-5	Stay put but may only fire canister next turn.
6	Fire as normal next turn.

Retreating Casualties

Square formations and column of march do not cause any further casualties. Cavalry who charged less than a standard move distance cause 1D3 extra casualties while cavalry who charged more than a standard move distance cause 1D6 extra casualties. Stationary cavalry cause no extra casualties. Retreating Cuirassiers equipped with back as well as breast plates ignore the first extra hit.

Close Combat

The number of dice each unit should roll is determined by its formation and whether the unit is facing the enemy to its front, flanks or rear. This is shown in the chart below:

	Facing Target Unit's:		
Unit's Formation	**Front**	**Flank**	**Rear**
Artillery Battery	1	1	1
Light Cavalry Line	3	1	1
Heavy Cavalry Line	3(4)	1	-
Cavalry Column	2/2	1	1
Infantry Column	2/2	1	1
Infantry Square	2	2	2
Infantry Line	3(6)	1	-
Skirmishers	1	1	1

Roll the appropriate number of dice and check the score of each die rolled on the chart below. This indicates the **minimum** score needed to register a hit on the enemy unit. Each successful hit causes one casualty.

Unit	Score Required to Hit
Infantry	4+
Heavy Cavalry	3+
Light Cavalry	4+
Skirmishers	6+
Artillery Crews	6+
Guards & Militia	
Guard Infantry	3+
Guard Heavy Cavalry	2+
Guard Light Cavalry	3+
Militia Infantry	5+

Retreating Casualties

Infantry

Type of Formation	Number of Extra Casualties Inflicted
Charging Column	1D6
Stationary Column	1D3
Charging Line	1D3
Stationary Line	None

Appendices

Appendix 1 – Base Sizes

These rules have been organised around the standard Foundry Base System of 40mm x 40mm bases for infantry (40mm x 50mm for cavalry). If you have already collected and painted armies and based them using a different system, this should make no difference to the game as long as all the participating armies are using the same basing system.

Many gamers will already own armies based according to the 'traditional' wargames system of six models on a 45mm x 40mm base, and as long as both armies are similarly based (even a few millimetres difference between base sizes won't make a crucial difference) then you'll be fine.

You can always reduce the size of your larger existing units or play with more models per unit than these rules recommend. As firing measurements are always made to the centre of a unit the overall size is not important. Larger cavalry units can be split down and used as multiple smaller units. This is particularly apt for French Guard units who had many squadrons, so for these rules you could quite easily field three units of Chasseurs a Cheval of the Guard.

Appendix 2 - Suggestions for Re-fighting Historical Battles

Many gamers have never tried, or never wished, to re-fight historical battles. The results can be very disappointing, especially when events follow the expected pattern of recorded history; after all, the French army was supposed to be surprised when the Prussians arrived at Waterloo! Re-fighting history can result in greatly advantaged or disadvantaged armies. It can take the pleasure out of gaming when you know your side has no chance of winning.

If you look at historical re-fights from another angle, though, they can be exhilarating, as you get to command very powerful or colourful armies and might be forced to fight on complex terrain. Scaling down the armies is one way to tackle historical battles. When recreating Lutzen in 1813 for example, use each division to represent an army corps. An army of three to four divisions will then consist of three or four corps. The Army Lists have been devised to make such changes in scale possible.

For smaller battles each division could directly represent an actual division or even a brigade, and with some modification a regiment can be made up of a number of battalions.

Unless your collection is vast and your table the size of a small hall, you could perhaps choose to focus on one section of a larger battle. Remember that in many Napoleonic battles armies deployed with a lot of depth, so you could look in more detail at the arrival of reserve troops or the instance of a flank attack.

Napoleon returns from exile, 1815, "Here I am. Kill your Emperor, if you wish." The soldiers responded with, "Vive L'Empereur!".

French Grenadier a Cheval.

We should always go before our enemies with confidence; otherwise our apparent uneasiness inspires them with greater boldness.

— Napoleon Bonaparte

French Mameluke (Knötel).

APPENDIX 3 - THE CAMPAIGNS OF GLORY

These troops have gained their experience in fifteen years of fighting.
- Friedrich Wilhelm III, King of Prussia

A storm was brewing, as a powerful westerly wind blew, dark thunder clouds were closing in, and once again the weather was becoming overcast over the English Channel as the despised British called it. A French Marshal with a voluminous blue great coat, which blew in the gentle wind, mounted his liver chestnut stallion and checked the tension on his girth. With a final adjustment, shortening the length of his gleaming stirrup leathers, he took off his Bicorne. With his receding dark hair rustling in the restless air, he took a handkerchief from an orderly standing to his right and wiped the humidity from his thick round spectacles, he replaced his hat upon his head and took a breath of sea air for the last time.

He spurred on his mount, pushing up a gentle grass slope, turned to four smartly dressed officers, his generals Bisson, Friant, Gudin and a lavishly attired cavalryman, Vialannes. He spoke "Gentlemen, this is the day that we leave the Armée des cotes de l'ocean is no more, we are now la Grande Armée, we are to head south our destination is the Rheinish city of Mannheim".

The warriors looked over their shoulders, a horse whinnied, musket barrels of his twenty-six thousand men gleamed in the darkening sky, they were about to leave the wooden huts which had been their home, behind was a broken *Peniche* landing craft wrecked upon the shore. The invasion of the red-coated English was no longer the premier goal, now it was back to fighting the white-coated Austrians.

It was only during last year's Spring that he had heard rumours of a French Empire was in the offing, with heredity, an Emperor and succession. He had given it his wholehearted support. Napoleon Bonaparte would no longer be first consul, but Emperor of France. Marshal Davoust was commander of the Third Corps, the best body of soldiers since Caesar's 10th legion. Davoust had been given his Marshals baton, once he was a simple Burgundian nobleman, he had to change his name, and became a great supporter of the Revolution, and the man he owed much of his promotion to, Napoleon Bonaparte.

Now on 27th August his Corps began to leave the camp de Bruges, little did he or any body else dream that within the next few years France, under their newly appointed Emperor would rule the greatest Empire since Roman times, stretching all across most of Europe.

This section allows you to collect and fight battles taking place within the great campaigns of 1805-1808. As the allies fought in a more traditional way, resembling the armies of the Seven Years War (1756–1763), you get a very different style of battles which require the skills of perhaps more experienced gamers to avoid them becoming to one sided.

THE FORMATION OF THE THIRD COALITION

On the 14th March 1804 a troop of French Dragoons commanded by General de Brigade, Michel Ordener

Early French infantry wearing Bicorne Hats, it's often forgotten that Napoleon's heroes of Austerlitz and Jena did not wear shakos.

captured the Bourbon related prince Duc d'Enghien, in the neutral territory of Baden. This Bourbon noble was then taken to Chateau de Vincennes on the outskirts of the French capital; court marshalled and then shot for treason. Austria, Prussia and Russia were outraged by this. On the 11th April 1805, the Third Coalition against France was forged; William Pitt the younger was the leading conspirator. A mighty alliance was formed between Britain, Russia and Austria; all they had to do was to finally convince Prussia (with her legendary invincible army built by Frederick the Great in the previous century) to join this crusade against Frances military dictator.

This was the excuse the allies had been waiting for ever' since Napoleon had placed a Crown upon his head on the 2nd October 1804 at the magnificent Notre Dame cathedral. The young French Republic had grown from, a citizen Levee en Masse into an Empire. Napoleon's further act of defiance against the old monarchies was his self coronation as King of Italy. What had already happened in France could happen to any of the other great European dynasties.

After the recent Revolutionary Wars, there had been a change in the style of warfare in the French armies from the Seven Years War. The Citizen armies of the French Republic had won battles, especially when lead by talented generals such as Messena, Hoche or Bonaparte himself. The battles of the Revolutionary Wars demonstrated that an inspired civilian army, fighting for a cause they believed in could match a traditional professional army.

The great campaigns of French Glory commenced with the Napoleons La Grande Armée departing from its operational base on the northern coast of France. This modern and mobile army then defeated an Austrian army under General Mack at the strategic battle of Ulm, simply by a series of marches. What then followed was perhaps Napoleon's greatest victory, crushing an Austro-Russian army at the Battle of Austerlitz. What was different about these battles was not only did it highlight the strength of a more modern Napoleonic Army, but it was also how this new style of army could obliterate its adversaries; two of Europe's major powers crushed in only a few weeks. – It would take Austria four years before her Army was able renew its struggle.

The following October the once thought of invincible army of Prussia, which Frederick the Great had created, was about to come into contact with Napoleon's war machine. The results were even more spectacular; in one day at the twin battles of Jena-Auestaedt Napoleon's troops reshaped Europe. Napoleon fought and defeated a smaller Prussian

army at Jena, and Marshal Davoust defeated a superior army with only a single French corps! The young and agile soldiers of France, who were not only good shots, but who were also able to perform deadly bayonet charges when required, had changed the style of warfare for ever.

The remnants of the defeated Prussians fled east and joined up with a Russian army as part of the Fourth Coalition. In the savage winter weather Napoleons army suffered its first halt, his army was too scattered and after two bloody days of inconclusive fighting, Napoleon was left in control of the battlefield but had failed to crush his adversaries.

Napoleon renewed the struggle on the 14th June, after twenty three hours of fighting at the battle of Friedland he decisively defeated Count Von Bennigsen's Russian army. Tsar Alexander I sued for peace, and met with Napoleon and signed the famous treaty of Tilsit. Napoleon Bonaparte was now master of Europe!

Early French infantry unit wearing Bicornes.

La Grande Armée

My enemies make appointments at my tomb.

— Napoleon Bonaparte

This period saw the Napoleon's army at its peak, capable of marching with unprecedented speed organised into formal divisions and corps, with reserves of Cuirassier cavalry and elite grenadier divisions taken from Battalions companies remaining in France. To complete this army, the Imperial Guard was building in strength, including veterans of Napoleons earlier victories. The army was incredibly powerful and tactically able to deploy in a way which perplexed her adversaries. Whilst this was the most modern army in Europe at that time, it is often forgotten that the infantry still wore bicorne hats and the longer tailed coats plus the army was still was mostly made up of Frenchmen!

French Army Special Rules 1805-07

The French army had been training at its camp in Northern France awaiting invasion of England (there is a theory that Napoleon never really intended to invade, but trained his army away from the prying eyes of his enemies) and all of the special rules in the Peninsular war section apply, with the following optional upgrades,

All line infantry may be chosen as line infantry or upgraded to elite infantry.

All light infantry may be chosen as light infantry or upgraded to elite infantry.

Foot dragoons are treated as elite line infantry, and may only be used against a Prussian or Prussian and Saxon army. Guard cavalry may not be used in this campaign.

Your army may include up to one division German allies against an Austrian-Russian Armies.

Your army may include up to two divisions of Bavarians against Prussian or Prussian and Saxon armies. Use Bavarian army lists.

Your army may include up to two divisions of Bavarians or other German allies against Russian or Russian and Prussian Armies. Use Bavarian army lists or if you wish to add any different German units, which are not in the Bavarian list, use the French list, provided your opponent agrees. Your army may also include up to two divisions of Saxons, use the Saxony list.

The Campaigns of Glory (1805-07) French

Divisions, Brigades & Batteries

Your army may include:

Type	Availability	Consisting of:
Infantry Division	Your army may include any number of infantry divisions.	3-5 infantry units, plus a divisional general; up to two units may be light infantry. Units may be of line or elite morale status.
Cavalry Brigade	Your army may include one cavalry brigade for each infantry division in the army.	1-2 units of cavalry; both units must be of the same type, except for cuirassiers and carabineer units which may be combined. At least half the cavalry units in the army must be light cavalry.
Cavalry Division	Two cavalry brigades may be combined into a cavalry division. You must meet the requirements for fielding both brigades.	As brigades; both brigades must be of the same type, with the exception of cuirassiers and carabineer units which may be mixed within a division. One brigade commander is promoted to divisional general.
Artillery Battery	Your army may include one artillery battery for every four infantry units it contains. If your army contains three infantry units or less, you may include a half battery at half the points cost.	The first artillery battery in your army must be a foot artillery battery, consisting of 3 foot artillery guns & 1 howitzer. Further artillery batteries may be either foot artillery batteries, or heavy artillery batteries consisting of 3 heavy artillery guns & 1 howitzer.

Horse Artillery Battery	Your army may include one horse artillery battery for every cavalry division it contains, or four every four units of cavalry which are not part of a division.	3 horse artillery guns & 1 howitzer.
Guard Division	Your army may include a single guard division if it contains at least three infantry divisions.	2-4 Old Guard infantry units, plus an Old Guard divisional general.
Combined Grenadier Division	Your army may include a single grenadier division if it contains at least three infantry divisions.	2-4 grenadier infantry units, plus a grenadier divisional general.
Foot Dragoon Division	Your army may include a single foot dragoon division if it contains at least three infantry divisions.	2-4 foot dragoon infantry units, plus a foot dragoon divisional general.
Guard Cavalry Brigade	Your army may include a single guard cavalry brigade if it includes a guard division or at least three infantry divisions. If your army includes two or more guard divisions, you may include two guard cavalry brigades.	The brigade may be either light or heavy. Guard light cavalry brigades consists of 1-3 units of guard light cavalry; guard heavy cavalry brigades consist of 1-2 units of guard heavy cavalry, plus a divisional general (not a brigade commander) in both cases.
Guard Artillery	You may upgrade artillery and horse artillery batteries to guard artillery if you army includes a guard infantry or cavalry division.	This is an upgrade and does not bring your army any additional units: any number of artillery batteries or horse artillery batteries can be upgraded to guard status.

Units

Choose the units that make up your divisions from the following:

Troop Type	Points Cost	Unit Size	Morale/Class
Line Infantry	20 points per unit	6	Line Infantry
Light Infantry	25 points per unit	6	Light Infantry
Elite Infantry	25 points per unit	6	Elite Infantry
Elite Light Infantry	30 points per unit	6	Elite Light Infantry
Foot Dragoons	30 points per unit	6	Elite Infantry
Allied Infantry	15 points per unit	6	Line Infantry
Old Guard Infantry	40 points per unit	6	Old Guard Infantry
Light Cavalry	35 points per unit	6	Light Cavalry
Dragoons	40 points per unit	8	Heavy Cavalry
Cuirassiers	45 points per unit	8	Heavy Cavalry
Carabineers	45 points per unit	8	Heavy Cavalry (no Cuirass)
Guard Light Cavalry	50 points per unit	6	Guard Light Cavalry
Guard Heavy Cavalry	55 points per unit	8	Guard Heavy Cavalry
Foot Artillery Battery	60 points/30 points	3 foot artillery & 1 howitzer	Foot Artillery
Heavy Artillery Battery	75 points	3 heavy artillery & 1 howitzer	Heavy Artillery
Horse Artillery Battery	75 points	2 horse artillery & 1 howitzer	Horse Artillery
Guard Foot Artillery	+10 points	-	Guard Foot Artillery
Guard Heavy Artillery	+10 points	-	Guard Heavy Artillery
Guard Horse Artillery	+10 points	-	Guard Horse Artillery

THE AUSTRIAN ARMY 1805

If you had seen one day of war, you would pray to God that you would never see another.
— Napoleon Bonaparte

The Austrian army had recently undergone extensive reorganisation; during the Revolutionary wars a number of very unsuccessful wars had almost bankrupt Austria! Many units had been disbanded, wagon trains and artillery teams had been broken up – all these measures were in a desperate attempt to save money.

The infantry were recruited from all over the extensive Habsburg Empire. General Mack who was awestruck by Napoleons campaigns and organisation had been able to quickly reorganise what had once been a bit of a hodgepodge of units, each organised by their own commanding officers, standardised into Battalions of Line Infantry and Grenadiers. However all this did was add to the confusion as the officers weren't used to the new structure. Even the muskets were a design relic dating from over fifty years ago, and no match for the improvements which had been made by French. No standard corps d'Armée organisation existed and columns of units were reshuffled on an almost weekly basis; this was confounded further by an army that drilled and fought in out dated linear formations.

The once proud Austrian Artillery was once again very antiquated. Again a relic of the previous century with battalion guns attached to infantry units there was no great attempt to concentrate artillery batteries on the battlefield. What confounded the artillery further was a lack of a modern limber teams, resulting in redeployment being slow at best.

Despite its backward looking deployment and tactics, the army was still capable of fighting well, and it took a general of Napoleon's genius to destroy it.

AUSTRIAN SPECIAL RULES

The larger size Austrian units are explained earlier in this book, but the following rules also apply to the early Austrian army.

Your Artillery may not form into Grand Batteries.

Your Infantry and cavalry units cannot deploy in columns of attack formation.

Despite the larger sized units (eight to ten stands) German line infantry roll 6d6 when firing a volley when deployed in line formation and Hungarian line infantry roll 8d6 when firing a volley when deployed in line formation.

No units in the army can perform forced march moves, unless the entire move is carried out along a road.

Artillery batteries (not including battalion guns) cannot perform forced marches, and can only make march moves on roads.

To simplify organisation, the columns of units are still classified as Divisions.

Austrian High Command.

Line infantry and grenadier units all have one attached battalion cannon.

Your army may ally with Russian Divisions, your cavalry and artillery entitlement is taken from each army on an individual basis.

Austrian Hussar Light Cavalry Unit.

Divisions, Brigades & Batteries

Your army may include:

Type	Availability	Consisting of:
Infantry Division	Your army may include any number of infantry divisions.	2-5 infantry units, and 0-1 detachment of Jaegers, plus a divisional general. If your army contains only a single infantry division, one unit may be grenadiers and one unit may be grenzers, otherwise all units must be fusiliers, Hungarian line infantry or militia. No individual unit type may exceed the number of fusilier units in the division.
Grenadier Division	Your army may include one grenadier division for every two infantry divisions it contains.	2-4 grenadier units, but plus a divisional general.
Advance Guard Division	Your army may include one advance guard division.	1-2 light cavalry units, 1 light infantry unit (of your choice), 1 Jaeger detachment, 1-2 Grenzer units and 1 horse artillery battery.
Light Cavalry Brigade	Your army may include one light cavalry brigade for every infantry division it contains.	1-2 light cavalry units, plus a brigade commander; one of these may be lancers.
Light Cavalry Division	Two light cavalry brigades may be combined to form a division; you must meet the requirements to field both brigades.	As brigades; one brigade commander is promoted to divisional general.
Dragoon Brigade	Your army may include a single dragoon brigade if it contains at least three infantry divisions.	1-2 dragoon units, plus a brigade commander.
Cuirassier Division	Your army may include one cuirassier division for every grenadier division it contains.	Two brigades, each consisting of 1-2 cuirassier or dragoon unit. One brigade has a brigade commander, the other has a divisional general. The division may also include a horse artillery battery.
Foot Artillery Battery	Your army may include one foot artillery battery for every six infantry units (not counting Jaegers) it contains, up to a maximum of one battery per infantry division.	3 foot artillery guns & 1 howitzer.
Heavy Artillery Battery	Your army may include one heavy artillery battery for every grenadier division of three or more units it contains.	2 heavy artillery guns & 1 howitzer.

Units

Choose the units that make up your divisions from the following:

Troop Type	Points Cost	Unit Size	Morale/Class
Fusiliers	25 points per unit	8 + 1 Battalion Gun.	Line Infantry
Grenadiers	35 points per unit	6 + 1 Battalion Gun.	Elite Infantry
Hungarian Line Infantry	30 points per unit	10 + 1 Battalion Gun.	Line Infantry
Grenzers	20 points per unit	6	Militia Infantry
Hussars	35 points per unit	6	Light Cavalry
Lancers	40 points per unit	6	Light Cavalry
Chevaux-Legers	40 points per unit	6	Light Cavalry
Dragoons	40 points per unit	8	Heavy Cavalry
Cuirassiers	45 points per unit	8	Heavy Cavalry
Jaegers	10 points per detachment	3	Skirmishers
Foot Artillery	60 points	3 foot & 1 howitzer	Artillery
Heavy Artillery	60 points	2 heavy & 1 howitzer	Artillery
Horse Artillery	60 points	3 horse	Artillery

SAXONY ARMY 1806-09

An army which cannot be regularly recruited is a doomed army. .

— Napoleon Bonaparte

Early Saxon Musketeers

In 1806 Friedrich August the Elector of Saxony allied to Prussia in an uneasy alliance. After the great disaster of Jena-Auestaedt in 1806, the Saxony army changed sides for the following seven years and fought for Napoleon. The Saxon army was organised along similar lines to the Prussian army with infantry regiments consisting of Musketeer and Grenadier units (battalions), although no units had battalion guns. The Saxon army was dressed in more French style uniform, of white coats, the artillery in green coats and the light cavalry in red coats; a very colourful army indeed!

SAXONY 1806-09 SPECIAL RULES

Infantry and cavalry units cannot deploy in columns of attack formation.

Artillery cannot form into Grand Batteries.

DIVISIONS, BATTERIES

Your army may include:

Type	Availability	Consisting of:
Infantry Division	Your army may include any number of infantry divisions.	3-5 infantry units, plus a divisional general. Up to one third may be Grenadiers.
Cavalry Brigade	Your army may include one cavalry brigade for each infantry divisions it contains.	1-2 units of Hussars or Chevauxlegers. If your army includes at least two infantry divisions, one cavalry brigade may consist of 1-2 units of heavy cavalry. If your army includes at least three infantry divisions, one cavalry brigade may consist of 1-2 units of cuirassiers or Garde du Corps.
Foot Artillery Battery	Your army may include on foot artillery battery for every four infantry units it contains.	3 foot artillery guns & 1 howitzer. Half battery consists of 2 foot artillery guns.
Horse Artillery Battery	Your army may include one horse artillery battery for every four cavalry units it contains (excluding cavalry squadrons).	3 horse artillery guns & 1 howitzer.

UNITS

Choose the units that make up your divisions from the following:

Troop Type	Points Cost	Unit Size	Morale/Class
Line Infantry	30 points per unit	6	Elite Infantry
Grenadiers	35 points per unit	6	Elite Infantry
Fusiliers	30 points per unit	6	Elite Light Infantry
Jaegers	35 points per unit	6	Elite Light Infantry
Hussars	35 points per unit	6	Light Cavalry
Cuirassiers (no armour)	45 points per unit	8	Heavy Cavalry
Garde du Corps	55 points per unit	8	Guard Cavalry
Foot Artillery	60 points	3 foot, 1 howitzer	Artillery

THE RUSSIAN ARMY 1805-08

In war, moral factors account for three quarters of the whole; material strength accounts for only one quarter.
— Napoleon Bonaparte

The army was organised into inspections, or inspectorates, each having a designated facing colour, red, turquoise, light The army was organised into inspections, or inspectorates, each having a designated facing colour, red, turquoise, light green, pink, lilac, blue, orange, white, yellow, ochre, raspberry, buff and grey; this means your army becomes a varied and colourful collection. No formal divisions were structured until the later part of the early period, when they were replaced with organised numbered divisions.

The line Infantry regiments contained musketeer and grenadier battalions (units) and, just to add a little more to the confusion, each musketeer battalion also contained a single grenadier company (and grenadier battalions contained fusilier and grenadier companies).

Light infantry units which made up about one third of the army; light infantry were made up of chasseur and carabineer companies.

Prussian style linear tactics with emphasis placed upon fire fighting was favoured, although some generals like Suvarov had favoured a bayonet attack in columns.

The Guard Units were selected from the strongest and best conscripts, which gave the army a solid reserve that could match up to any thing Napoleon could attack them with.

The cavalry was also supplemented by many irregulars and as always in the Russian artillery was of excellent quality. For all its size and theatrical power the army had not yet become involved in a great popular crusade and with the battles taking place outside of the homeland the army was not quite as resolute as it would became in its later campaigns.

RUSSIAN ARMY 1805-08 SPECIAL RULES

As already mentioned although the army could and did form columns of attack, line formation is the preferred battle deployment for infantry. Only infantry units led by a division general with a command rating of seven or eight can use the column of attack formation, those lower representing the innate Russian conservatism!

None of the Russian special rules found earlier in this book for the later army lists apply.

The army is organised into Columns which to avoid confusion are still classified in the rules as divisions, (which are a bit like a large divisions or a small corps so count army design is based around).

Artillery cannot form into Grand Batteries.

Your army may ally with Austrian Divisions or Prussian (which may not exceed four three divisions), your cavalry and artillery entitlement is taken from each army on an individual basis.

Early Russian Horse Artillery.

Early Russian Musketeer unit.

Divisions, Brigades & Batteries

Your army may include:

Type	Availability	Consisting of:
Infantry Division	Your army may include any number of infantry divisions.	3-6 infantry units, plus a divisional general; up to a third of the units in each division may be Jaegers and or Grenadiers. Each division includes a pulk of Cossacks.
Grenadier Division	Your army may include one grenadier division for every two infantry divisions it contains.	3-4 grenadiers units, plus a divisional commander.
Guard Division	Your army may include a single guard division if it contains at least three infantry divisions.	2-4 guard infantry units, one unit may be Guard Jaeger, plus a divisional commander.
Guard Cavalry Brigade	Your army may include up to two guard cavalry brigades if it includes a guard infantry division.	1-2 guard cavalry units, plus a brigade commander.
Guard Cavalry Division	Two guard cavalry brigades may be combined to form a guard cavalry division; you must meet the requirements for fielding both brigades.	As brigades; one brigade commander is promoted to divisional general.
Light Cavalry Brigade	Your army may include one light cavalry brigade for each infantry division it contains.	1-2 light cavalry units, plus a brigade commander.
Cuirassier Brigade	Your army may include a single cuirassier brigade if it contains a grenadier division.	1-2 cuirassier units, plus a brigade commander.
Cavalry Division	Your army may include one cavalry division for every two infantry divisions it contains; a cavalry division count as two brigade choices.	Two cavalry brigades of 1-2 units; one brigade may consist of heavy cavalry, the other must consist of light cavalry. Both units in a brigade must be of the same type. One brigade has a brigade commander; the other has a divisional general.
Cossack Brigade	Your army may include a single Cossack brigade if it contains at least three infantry divisions.	1-2 Cossack units, plus a divisional commander.
Irregular Cavalry Units	Your army may include one irregular cavalry unit for every cavalry division contains.	1 irregular cavalry unit.
Foot Artillery Battery	Your army may include one artillery battery for every five infantry units it contains. If your army includes four or less units of infantry, you may include a half battery at half the points cost.	4 foot artillery guns & 2 howitzers. Half battery consist of 2 foot artillery guns & 1 howitzer.
Heavy Artillery Battery	One foot artillery battery me upgraded to heavy artillery for every three foot artillery	This is an upgrade and does not provide the army with any new units: the heavy artillery battery replaces one foot artillery battery and consists of 4 heavy artillery guns & 2 howitzers.
Horse Artillery Battery	Your army may include one horse artillery battery for every four cavalry units in the army (not counting Cossacks or irregular cavalry). If your army contains three or less units of cavalry, you may include a half battery at half the points cost.	Either: 6 horse artillery guns, or 4 horse artillery guns & 1 howitzer. Half batter consists of either: 3 horse artillery guns, or 2 horse artillery guns & 1 howitzer.
Guard Artillery	One of your foot artillery batteries may be upgraded to guard artillery if your army includes a guard infantry division.	This is an upgrade and does not provide any additional units to the army: the upgraded battery becomes a guard artillery battery.
Guard Horse Artillery	One of your horse artillery batteries may be upgraded to guard horse artillery if your army includes a guard cavalry brigade.	This is an upgrade and does not provide any additional units to the army: the upgraded battery becomes a guard horse artillery battery.

UNITS

Choose the units that make up your divisions from the following:

Troop Type	Points Cost	Unit Size	Morale/Class
Line Infantry	15 points per unit	6	Line Infantry
Jaegers	20 points per unit	6	Light Infantry
Grenadiers	25 points per unit	6	Elite Infantry
Guard Infantry	30 points per unit	6	Guard Infantry
Guard Jaegers	35 points per unit	6	Guard Light Infantry
Hussars	35 points per unit	6	Light Cavalry
Lancers	40 points per unit	6	Light Cavalry
Regular Cossacks	40 points per unit	6	Light Cavalry
Irregular Cavalry	20 points per unit	6	Militia Cavalry
Dragoons	40 points per unit	8	Heavy Cavalry
Cuirassiers	45 points per unit	8	Heavy Cavalry
Guard Cossacks	55 points per unit	6	Guard Light Cavalry
Guard Dragoons	55 points per unit	8	Guard Heavy Cavalry
Guard Cuirassiers	60 points per unit	8	Guard Heavy Cavalry
Guard Hussars	50 points per unit	6	Guard Light Cavalry
Mounted Jaegers	35 points per unit	6	Light Cavalry
Foot Artillery Battery	90 points	4 foot & 2 howitzers	Artillery
Heavy Artillery Battery	120 points	4 heavy & 2 howitzers	(Positional) Artillery
Horse Artillery	100 points	6 horse, or 4 horse & 2 howitzers	Artillery
Guard Foot Artillery Battery	+15 points	-	Guard Artillery
Guard Horse Artillery Battery	+15 points	-	Guard Horse Artillery

The Battle of Austerlitz, December 2, 1805.

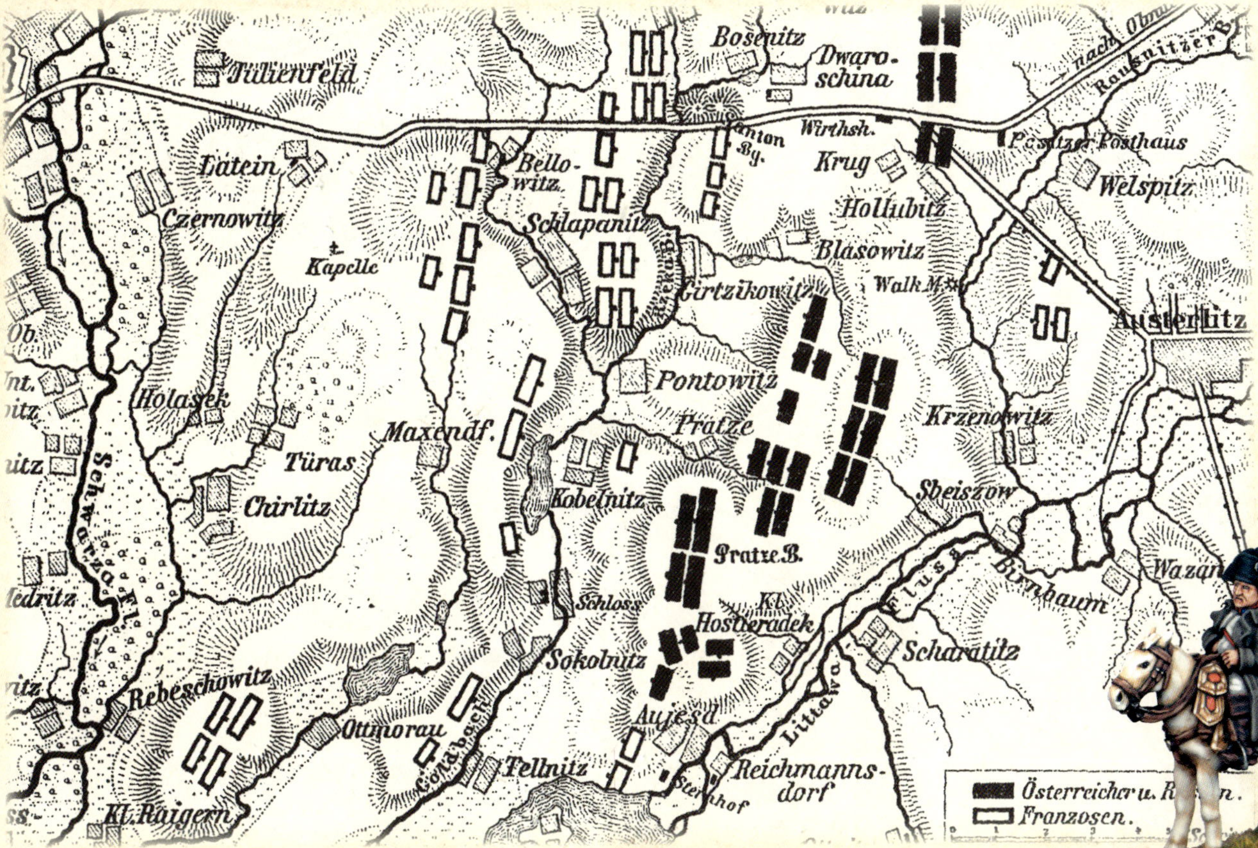

Boney.

The Prussian Army 1806-07

A general-in-chief should ask himself several times in the day, "What if the enemy were to appear now in my front, or on my right, or my left?".

— Napoleon Bonaparte

The shadow of Old Fritz (Frederick the Great) stalked throughout the army, the memory of his ability and the greatness of his armies invincibility gave the soldiers of Prussia an over zealous confidence. The French army, with its new ways of fighting may have crushed the and Austrian armies, driven the warriors of Holy Russia back to their homeland but the descendants of the heroes of the Seven Years War would prove once and for all the old ways of fighting were still the best.

The Prussian cavalry, many of whom had famously sharpened their swords on the French Embassy in Berlin, were the best trained and mounted on the best horses in Europe. The infantry units comprised of Musketeer (line), grenadier and Fusilier (light), supplemented by a crack Jager force. The artillery was still partially split in the traditional 18th century way, with battalion guns attached to Musketeer and Grenadier units; linear tactics were hung on to, even after their unsuccessful use by the Austrian and Russians.

Prussian army 1806-08 Special Rules

None of the Prussian special rules found earlier in this book for the later army lists apply.

The Prussian army is either of elite or Guard Status!

Infantry and cavalry units cannot deploy in columns of attack formation.

Grenadiers roll 7d6 when firing volleys when deployed in line formation.

All Line and grenadier have an attached battalion cannon.

Artillery cannot form into Grand Batteries.

Your Prussian army may ally with Saxon Divisions, the cavalry and artillery entitlement is taken from each army on an individual basis.

Early Prussian Musketeer unit wearing Bicornes.

Early Prussian Musketeer battalion gun & crew.

Early Prussian Grenadier unit.

DIVISIONS, BATTERIES

Your army may include:

Type	Availability	Consisting of:
Infantry Division	Your army may include any number of infantry divisions.	3-5 infantry units, plus a divisional general. Up to one third of the Fusiliers may be line infantry, up to one third Grenadier.
Cavalry Brigade	Your army may include one cavalry brigade for each infantry divisions it contains.	1-2 units of Hussars or one Lancer. If your army includes at least two infantry brigades, one cavalry brigade may consist of 1-2 units of dragoons. If your army includes at least three infantry brigades, one cavalry brigade may consist of 1-2 units of cuirassiers.
Foot Artillery Battery	Your army may include on foot artillery battery for every six infantry units it contains.	3 foot artillery guns & 1 howitzer. Half battery consists of 2 foot artillery guns. If your army contains three or more infantry brigades, one battery may be a heavy artillery battery consisting of 3 heavy artillery guns & 1 howitzer.
Horse Artillery Battery	Your army may include one horse artillery battery for every four cavalry units it contains.	3 horse artillery guns & 1 howitzer.
Guards	You may upgrade two line infantry units and or Grenadiers in a single division to Guard status, if your army contains three or more infantry divisions.	This is an upgrade and does not provide any additional units to the army.

UNITS

Choose the units that make up your divisions from the following:

Troop Type	Points Cost	Unit Size	Morale/Class
Line Infantry (Musketeers)	30 points per unit	6+1 Battalion Gun	Elite Infantry
Grenadiers	35 points per unit	6+1 Battalion Gun	Elite Infantry
Guards	+20 points per unit	-	Guard Infantry
Fusiliers	30 points per unit	6	Elite Light Infantry
Jaegers	35 points per unit	6	Elite Light Infantry
Hussars	35 points per unit	6	Elite Light Cavalry
Lancers	40 points per unit	6	Elite Light Cavalry
Dragoons	50 points per unit	8	Elite Heavy Cavalry
Cuirassiers (no armour)	50 points per unit	8	Elite Heavy Cavalry
Foot Artillery	60 points	3 foot, 1 howitzer	Artillery
Heavy Artillery (positional)	75 points	3 foot, 1 howitzer	Artillery
Horse Artillery	75 points	3 horse, 1 howitzer	Artillery

Early Prussian Fusiliers unit.

INDEX

French Engineer (Knötel).

INDEX

French Guard Chasseur (Knötel).

INDEX

French Grenadier (Knötel).

TAKE YOUR PAINTING TO THE NEXT LEVEL!

KEVIN DALLIMORE'S FULLY COMPREHENSIVE GUIDE TO PAINTING & BUILDING MINIATURE MODELS.

If you're entirely new to painting then this guide will help to get you started, and if you have painted before you'll find it full of useful tips and new ideas.

With the aid of fully illustrated stage-by-stage examples Kevin describes every aspect of his celebrated painting style, with a quick and simple method designed to get you painting single models and armies to a high standard.

All areas of miniature model painting are examined in detail, with sections covering the equipment you'll need, the preparation of models, designing and painting shields, using transfers, painting horses using oils, varnishing and basing. There are also features on creating dioramas, converting models and painting vehicles and buildings plus hints and tips from some well-known guest painters.

Illustrated throughout with a vast number of painted examples featuring Kevin's stunning work, this guide should inspire miniature model enthusiasts everywhere to pick up those paint brushes!

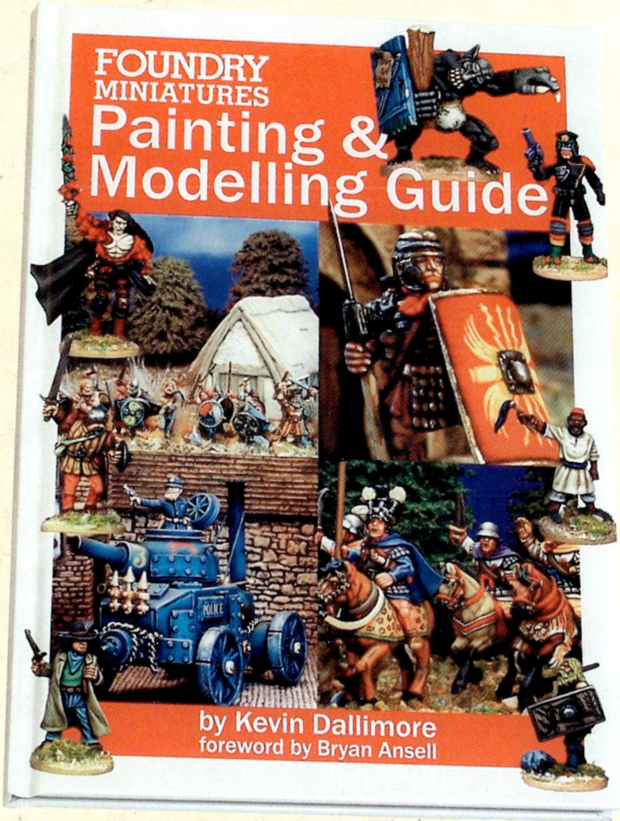

" If you paint wargames-scale model soldiers you need to read this book"

Chris Steadman

" Looks fantastic. Full-colour Eye Candy everywhere you look"

Steve Dean

THE FOUNDRY PAINT SYSTEM

NAPOLEONIC AUTHENTIC COLOURS

SHADE 65A	FRENCH BLUE 65B	LIGHT 65A
SHADE 69A	RUSSIAN GREEN 69B	LIGHT 69A
SHADE 73A	BAVARIAN CORNFLOWER BLUE 73B	LIGHT 73A
SHADE 66A	PRUSSIAN BLUE 66B	LIGHT 66A
SHADE 70A	FRENCH DRAGOON GREEN 70B	LIGHT 70A
SHADE 74A	BRITISH ROYAL BLUE 74B	LIGHT 74A
SHADE 67A	AUSTRIAN WHITE 67B	LIGHT 67A
SHADE 71A	FRENCH CHASSEUR A CHEVAL GREEN 71B	LIGHT 71A
SHADE 75A	BRITISH BLUE GREY 75B	LIGHT 75A
SHADE 68A	BRITISH REDCOAT 68B	LIGHT 68A
SHADE 72A	MUSKET STOCK BROWN 72B	LIGHT 72A
SHADE 75A	FRENCH HUSSAR SKY BLUE 75B	LIGHT 75A

36 COLOURS IN 12 MATCHED SETS, PLUS 6 BRILLIANT BRUSHES

Since launching the Foundry Paint System, inspired by Kevin Dallimore's three-colour painting technique, Foundry has received many queries regarding which colours are best to use for specific historical uniforms. As most of these enquiries are directed at our huge range of Napoleonic models ("What blue should I use for French line infantry coats?" being a very common question) we decided it was time to design some paints especially for reproducing the uniforms of this popular period.

In reality the exact shades of uniform colours varied, often within the same unit! With this in mind we have designed our paints to give the best interpretation of uniform colours, while still producing attractive shades that compliment our existing range of paints. With our new colours your models will look equally stunning in the display cabinet or on the table top, with enough Regency splendour to grace the pages of a Jane Austen novel!

Although they are aimed at Napoleonic uniforms these colours are ideal for other historical periods. The British colours, for example, will suit the British army from the Seven Years War right up to the Great War. Whatever types of models you are painting our new colours make a great addition to the Foundry Paint System … You can never have enough blues, greens and browns!

Recreate the stunning appearance of the soldiers from the 'golden age' of military uniforms, including six fine quality brushes for painting all that fancy lace and trim! Get the 'British Soldier' model when you buy the NAPOLEONIC AUTHENTIC COLOURS Painting Set!

2 x FOUNDRY PURE SABLE FINE DETAIL size 0000 (for fine detail work)

1 x FOUNDRY PURE SABLE FINE FINISH size 000 (for detail work)

1 x FOUNDRY SABLE MIX GENERAL DETAIL size 00 (for larger detail work)

1 x FOUNDRY BEST NYLON BLOCKING size 1 (for blocking in larger areas and horses)

1 x FOUNDRY UTILITY DRY BRUSHING size 2 (nylon flat mixing).

THE FOUNDRY PAINT SYSTEM

THE FOUNDRY PAINT SYSTEM COMBINED SET: THE ORIGINAL SET PLUS THE EXPANSION SET

THE EXPANSION SET: COLOUR SETS 33, 34 & 37 TO 64

SHADE 53A	CHESTNUT 53B	LIGHT 53C
SHADE 54A	CONKER BROWN 54B	LIGHT 54C
SHADE 55A	BUTTER FUDGE 55B	LIGHT 55C
SHADE 56A	PALOMINO 56B	LIGHT 56C
SHADE 57A	STONE 57B	LIGHT 57C
SHADE 58A	DEEP MAUVE 58B	LIGHT 58C
SHADE 59A	RED BRICK 59B	LIGHT 59C
SHADE 60A	MADDER RED 60B	LIGHT 60C
SHADE 61A	PEATY BROWN 61B	LIGHT 61C
SHADE 62A	NIGHT SKY 62B	LIGHT 62C
SHADE 63A	QUAGMIRE 63B	LIGHT 63C
SHADE 64A	DEEP MAROON 64B	LIGHT 64C
COPPER WASH 47A	BROWN WASH 47B	BLACK WASH 47C
WHITE 33C	WHITE 33C	WHITE 33C
WHITE 33C	WHITE 33C	WHITE 33C
BLACK 34A	BLACK 34A	BLACK 34A
BLACK 34A	BLACK 34A	BLACK 34A
SHADE 37A	TERRACOTTA 37B	LIGHT 37C
SHADE 38A	SCARLET 38B	LIGHT 38C
SHADE 39A	STORM BLUE 39B	LIGHT 39C
SHADE 40A	VIOLET 40B	LIGHT 40C
SHADE 41A	BRIGHT BOTTLE GREEN 41B	LIGHT 41C
SHADE 42A	BAY BROWN 42B	LIGHT 42C
SHADE 43A	BURNISHED COPPER 43B	LIGHT 43C
SHADE 44A	BURNING GOLD 44B	LIGHT 44C
SHADE 45A	DEEP BROWN LEATHER 45B	LIGHT 45C
SHADE 46A	CADAVEROUS GREEN 46B	LIGHT 46C

Each main colour is teamed up with the highlight and shading colours that we would choose to use ourselves for painting a model, so that you quickly achieve results like those you see on our web, in our adverts and, of course, on the painted miniatures in this book.

Our paints really cover your models; we haven't joined the bandwagon and economised by cutting back on the amount of pigment in our paint. We believe that our paint contains the highest proportion of actual pigment of any paint available in the hobby market.

Warm, glowing colours with real depth, just like the ones we've been mixing ourselves for years.

20ml pots (everyone else seems to have reduced their contents to a little over half that) with lids that really provide an airtight seal.

THE ORIGINAL SET: COLOUR SETS 1 TO 36

SHADE 19A	ROYAL PURPLE 19B	LIGHT 19C
SHADE 20A	DEEP BLUE 20B	LIGHT 20C
SHADE 21A	SKY BLUE 21B	LIGHT 21C
SHADE 22A	VIVID BLUE 22B	LIGHT 22C
SHADE 23A	TOMB BLUE 23B	LIGHT 23C
SHADE 24A	TEAL BLUE 24B	LIGHT 24C
SHADE 25A	BRIGHT GREEN 25B	LIGHT 25C
SHADE 26A	FOREST GREEN 26B	LIGHT 26C
SHADE 27A	STORM GREEN 27B	LIGHT 27C
SHADE 28A	PHLEGM GREEN 28B	LIGHT 28C
SHADE 29A	MOSS 29B	LIGHT 29C
SHADE 30A	RAW LINEN 30B	LIGHT 30C
SHADE 31A	GRANITE 31B	LIGHT 31C
SHADE 32A	SLATE GREAY 32B	LIGHT 32C
SHADE 33A	ARCTIC GREY 33B	WHITE 33C
BLACK 34A	CHARCOAL BLACK 34B	LIGHT 34C
CHAINMAIL 35A	ARMOUR 35B	SPEARPOINT 35C
BRAZEN 36A	GOLD 36B	SHINY 36C
SHADE 1A	LEMON 1B	LIGHT 1C
SHADE 2A	YELLOW 2B	LIGHT 2C
SHADE 3A	ORANGE 3B	LIGHT 3C
SHADE 4A	OCHRE 4B	LIGHT 4C
SHADE 5A	FLESH 5B	LIGHT 5C
SHADE 6A	DUSKY FLESH 6B	LIGHT 6C
SHADE 7A	BUFF LEATHER 7B	LIGHT 7C
SHADE 8A	CANVAS 8B	LIGHT 8C
SHADE 9A	BONE YARD 9B	LIGHT 9C
SHADE 10A	BASE SAND 10B	LIGHT 11C
SHADE 11A	RAWHIDE 11B	LIGHT 11C
SHADE 12A	DRAB 12B	LIGHT 12C
SHADE 13A	SPEARSHAFT 13B	LIGHT 13C
SHADE 14A	TAN 14B	LIGHT 14C
SHADE 15A	BRIGHT RED 15B	LIGHT 15C
SHADE 16A	NIPPLE PINK 16B	LIGHT 16C
SHADE 17A	WINE STAIN RED 17B	LIGHT 17C
SHADE 18A	SHADOW PINK 18B	LIGHT 18C

1644

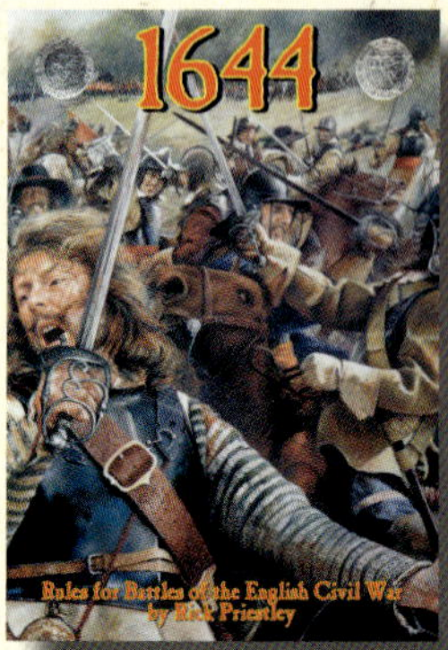

Rules for Battles of the English Civil War

With 1644 the reader is given a very enjoyable insight into the English Civil War, together with plenty of additional help and advice to enable either the experienced, or novice player to get the maximum enjoyment from playing an interesting and challenging game devised by one of the best writers in the wargames industry, Rick Priestley. 1644 also has painting sections written by Kevin Dallimore on how to paint a pikeman and flags and interesting background articles on the Bishops Wars, the English Civil War and the Thirty Years' War. 112 pages, lavishly illustrated in full colour. A4 Softback

Medieval Warfare

Rules for Medieval Battles 450-1515 AD

Terry's long awaited, definitive and detailed set of wargame rules for the Medieval period. An A4 hardback with 168 pages, lavishly illustrated throughout in black and white with colour sections. Special illustrations by Peter Denis with cover illustrations by Chris Collingwood. Spanning everything from the Late Romans to the beginning of the sixteenth century; with these rules you can command and fight against some of the most exciting armies ever to stride the battlefields of Europe and the Middle East. More than 80 colour illustrations, over 172 black and white illustrations, 36 diagrams and maps.

The Foundry Compendium

Pirates to Darkest Africa

The Foundry Compendium is packed with rules, painting advice and ideas for making great-looking terrain. It will appeal to anyone interested in playing games such as exploration in Darkest Africa, fighting for survival and glory in Rome's gladiatorial arenas and plundering the West African coast with your band of cutthroat pirates. Themed to concentrate on Pirates and Darkest Africa, but covering other historical periods too; Aztecs, Gladiators, Greeks and Romans. 96 pages, lavishly illustrated in full colour throughout. A4 Softback

The Rules With No Name

Rules for the Old West

The Rules With No Name are a straightforward set of rules for gunfights in the age of the Cowboy and the Indian, the Lawman and the Outlaw. Classic encounters are played out using 28mm miniatures in an easy-to-learn game full of subtle complexity. The innovative Fate Deck introduces a fun and realistic element of chance while simple rules for movement, shooting and fighting allow for quick, friendly games in which players will find themselves presented with a deceptively challenging set of tactical choices from one turn to the next. This book contains all the rules you will need to play. 136 pages, lavishly illustrated in full colour throughout.

The Paraguayan War

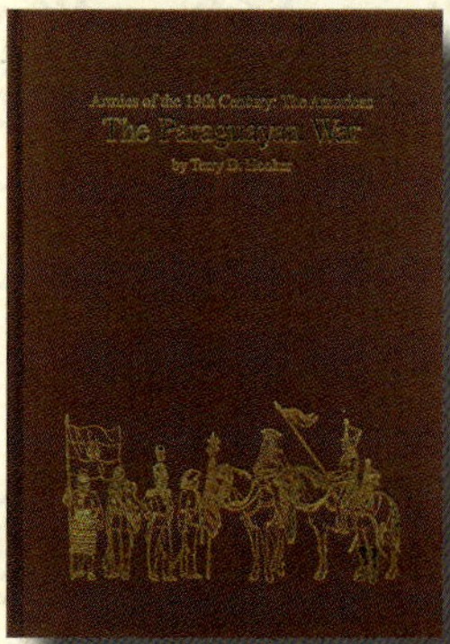

Organisation, Warfare, Dress and Weapons

The Paraguayan War, sometimes referred to as the Triple Alliance War, was a sanguinary conflict of attrition fought mostly within a triangle of land formed by the Parana, Terry Hooker provides not only a lucid account of the war itself – the largest ever fought in South America – but also an outline of the convoluted political and military events which led up to it. 192 pages. Illustrations include 252 figures, 60 illustrations, 16 maps.

Just one of a long series of definitive military reference books, with volumes covering the armies of a variety of periods and nations, mostly presented in an A4 sized hardback format with traditional linen and glittering gilt binding.

9me Régiment de Hussards, Cie d'Élite, 1812
Trompette